FEEDBACK CONTROL FOR PERSONALIZED MEDICINE

FEEDBACK CONTROL FOR PERSONALIZED MEDICINE

Edited by

Esteban A. Hernandez-Vargas
Institute of Mathematics, UNAM
Juriquilla, Mexico

ELSEVIER

ACADEMIC PRESS
An imprint of Elsevier

Academic Press is an imprint of Elsevier
125 London Wall, London EC2Y 5AS, United Kingdom
525 B Street, Suite 1650, San Diego, CA 92101, United States
50 Hampshire Street, 5th Floor, Cambridge, MA 02139, United States
The Boulevard, Langford Lane, Kidlington, Oxford OX5 1GB, United Kingdom

ISBN: 978-0-323-90171-0

For information on all Academic Press publications
visit our website at https://www.elsevier.com/books-and-journals

Publisher: Mara Conner
Acquisitions Editor: Sonnini R. Yura
Editorial Project Manager: Emily Thomson
Production Project Manager: Prem Kumar Kaliamoorthi
Designer: Mark Rogers

Typeset by VTeX

To the most certain in uncertain hours, God
and to my partner, friend, enemy, and wife, Marwa

Esteban A. Hernandez Vargas

Contents

Contributors xi
About the editor xiii
Preface xv
Acknowledgments xvii

1. Closing the loop in personalized medicine
Esteban A. Hernandez-Vargas

References 7

2. Optimal control strategies to tailor antivirals for acute infectious diseases in the host: a study case of COVID-19
Mara Pérez, Pablo Abuin, Marcelo Actis, Antonio Ferramosca, Esteban A. Hernandez-Vargas, and Alejandro H. González

2.1 Introduction 11
2.2 Review of the target cell limited model for in-host infection 13
2.3 Equilibrium characterization and stability 15
2.4 Inclusion of PK and PD of antiviral treatment 23
2.5 Control strategies to tailor therapies 26
2.6 Simulation results 32
2.7 Conclusions and future works 36
References 37

3. Input-output approaches for personalized drug dosing of antibiotics
Pauline Thémans, Flora Tshinanu Musuamba, and Joseph J. Winkin

3.1 Introduction 41
3.2 Population pharmacokinetic model 42
3.3 Individualized drug dosing 47
3.4 State estimation and feedback control law 59
3.5 Conclusion 63
Acknowledgments 64
References 64

4. Safe glycemia regulation considering parameter variations under the offset-free MPC with pulse inputs scheme
María F. Villa-Tamayo and Pablo S. Rivadeneira

4.1 Introduction 67
4.2 System with pulse inputs 69
4.3 Offset-free MPC strategy 72
4.4 Application to T1DM treatment 74
4.5 Safety layer for insulin-on-board constraint 78
4.6 Conclusion 84
References 84

5. Deep neuronal network-based glucose prediction for personalized medicine
Alma Y. Alanis, Oscar D. Sanchez, and Eduardo Ruiz Velázquez

5.1 Introduction 87
5.2 Deep neural networks 90
5.3 Direct multistep ahead prediction strategy 94
5.4 System description 95
5.5 Results 98
5.6 Conclusion and discussion 104
Acknowledgments 104
References 104

6. Control-based drug tailoring schemes towards personalized influenza treatment
Gustavo Hernandez-Mejia

6.1 Introduction 109
6.2 Influenza virus in the host 111
6.3 Influenza treatment and therapy optimization 114
6.4 Control schemes towards treatment tailoring 116
6.5 Inverse optimal impulsive control 118
6.6 Influenza treatment tailoring 120
6.7 Conclusions and final remarks 123
References 124

7. Polynomial state estimation in infectious diseases
M. Hernandez-Gonzalez

7.1 Introduction 129
7.2 Preliminaries 131
7.3 Statement of the problem 133
7.4 Solution of the problem 134
7.5 State estimation for systems of viral infections 139
7.6 Conclusions 146
References 147

8. Sliding mode control theory interprets elite control of HIV

Anet J.N. Anelone, Peter Kim, and Sarah K. Spurgeon

8.1 Introduction 151
8.2 Modeling HIV infection in vivo 154
8.3 A reachability condition for elite control of HIV infection 158
8.4 Robustness properties of elite control of HIV infection 160
8.5 Simulation results 162
8.6 Conclusion 169
Acknowledgments 169
References 169

9. A stochastic model for hepatitis C viral infection dynamics with the innate immune response

Mauricio Castaño-Arcila and Jesús Rodríguez-González

9.1 Introduction 173
9.2 Biological background 174
9.3 Stochastic model 178
9.4 Results 179
9.5 Conclusion 184
Conflict of interest statement 184
Author contributions 185
Funding 185
References 185

10. Impulsive nonlinear MPC with application to oncolytic virus therapy

María F. Villa-Tamayo, Anet J.N. Anelone, and Pablo S. Rivadeneira

10.1 Introduction 189
10.2 Notation 191
10.3 Model of oncolytic virus therapy 191
10.4 Model predictive control for oncolytic virus therapy 193
10.5 Results 196
10.6 Conclusion 198
References 200

11. Is the isolated heart a relaxation-oscillator?

Carlos Enrique Gálvez-de León, Amelia Ríos, Karen Garza Cuellar, Bruno A. Escalante, and Jesús Rodríguez-González

11.1 Introduction 203
11.2 Biological background 204
11.3 Theoretical background 205
11.4 Methodology 207

11.5 Results 209
11.6 Conclusion 211
Conflict of interest statement 216
Author contributions 216
Funding 216
References 216

Index 219

Contributors

Pablo Abuin Institute of Technological Development for the Chemical Industry (INTEC), CONICET-UNL, Santa Fe, Argentina

Marcelo Actis Facultad de Ingeniería Química (FIQ), Universidad Nacional del Litoral (UNL) and Consejo Nacional de Investigaciones científicas y técnicas (CONICET), Santa Fe, Argentina

Alma Y. Alanis University Center for Exact Sciences and Engineering, University of Guadalajara, Guadalajara, Jalisco, Mexico

Anet J.N. Anelone School of Mathematics and Statistics, University of Sydney, Camperdown, NSW, Australia

Mauricio Castaño-Arcila Francisco José de Caldas District University, Bogotá, Colombia

Bruno A. Escalante Center for Research and Advanced Studies of the National Polytechnic Institute, Monterrey Unit, Parque PIIT, Apodaca, NL, Mexico

Antonio Ferramosca Department of Management, Information and Production Engineering, University of Bergamo, Dalmine, BG, Italy

Carlos Enrique Gálvez-de León Center for Research and Advanced Studies of the National Polytechnic Institute, Monterrey Unit, Parque PIIT, Apodaca, NL, Mexico

Karen Garza Cuellar Center for Research and Advanced Studies of the National Polytechnic Institute, Monterrey Unit, Parque PIIT, Apodaca, NL, Mexico

Alejandro H. González Institute of Technological Development for the Chemical Industry (INTEC), CONICET-UNL, Santa Fe, Argentina

M. Hernandez-Gonzalez Autonomous University of Nuevo Leon, Department of Physical and Mathematical Sciences, San Nicolas de los Garza, Nuevo Leon, Mexico

Gustavo Hernandez-Mejia Institute of Epidemiology and Social Medicine, University of Münster, Münster, Germany

Esteban A. Hernandez-Vargas Institute of Mathematics, UNAM, Juriquilla, Mexico

Peter Kim School of Mathematics and Statistics, University of Sydney, Camperdown, NSW, Australia

Flora Tshinanu Musuamba Federal Agency for Medicines and Health Products, Brussels, Belgium

Mara Pérez Institute of Technological Development for the Chemical Industry (INTEC), CONICET-UNL, Santa Fe, Argentina

Amelia Ríos Center for Research and Advanced Studies of the National Polytechnic Institute, Monterrey Unit, Parque PIIT, Apodaca, NL, Mexico

Pablo S. Rivadeneira Universidad Nacional de Colombia, Facultad de Minas, Grupo GITA, Medellín, Colombia

Jesús Rodríguez-González Center for Research and Advanced Studies of the National Polytechnic Institute, Monterrey Unit, Parque PIIT, Apodaca, NL, Mexico

Oscar D. Sanchez University Center for Exact Sciences and Engineering, University of Guadalajara, Guadalajara, Jalisco, Mexico

Sarah K. Spurgeon Department of Electronic and Electrical Engineering, University College London, London, United Kingdom

Pauline Thémans Namur Institute for Complex Systems (naXys) and Department of Mathematics, University of Namur, Namur, Belgium

Eduardo Ruiz Velázquez University Center for Exact Sciences and Engineering, University of Guadalajara, Guadalajara, Jalisco, Mexico

María F. Villa-Tamayo Universidad Nacional de Colombia, Facultad de Minas, Grupo GITA, Medellín, Colombia

Joseph J. Winkin Namur Institute for Complex Systems (naXys) and Department of Mathematics, University of Namur, Namur, Belgium

About the editor

Esteban A. Hernandez-Vargas is a research group leader and principal investigator at the Institute of Mathematics, UNAM, Mexico. In 2011, he obtained his PhD in Mathematics from the Hamilton Institute at the National University of Ireland. During 3 years, he held a postdoctoral scientist position at the Helmholtz Centre for Infection Research (HZI) in Braunschweig, Germany. In July 2014, he founded the pioneering research group of Systems Medicine of Infectious Diseases at the HZI. In March 2017, he and his research group moved to the Frankfurt Institute for Advanced Studies. Furthermore, he was a visiting scholar at Los Alamos National Laboratory (USA), Universidad de Guadalajara (Mexico), CentraleSupelec (France), and Northeastern University (USA) and adjunct lecturer at the Otto von Guericke Universität Magdeburg, Germany. He has written more than 100 peer-reviewed publications. He is an Editorial Board member of *Plos One*, the *Journal of Franklin Institute*, *Viruses*, *Frontiers of Immunology*, and *Artificial Intelligence Review*. He is also involved in several IEEE and IFAC Conference Organizing Committees.

Preface

This book reflects recent developments in personalized medicine by the control theoretical community. This provides a comprehensive introduction to multidisciplinary fields like virology, immunology, epidemiology, statistics, machine learning, and control systems. The book here presents a set of 11 chapters introducing how control theoretical approaches can help to tailor therapies for different diseases, in particular, infectious diseases, diabetes, and cancer.

Basic knowledge of control theory is helpful though not essential as the book includes a self-contained introduction in each chapter. The book is mainly recommended for graduate students and scientists working in the control field with biomedical applications. In particular, this book provides an effort to couple control theoretical approaches, PK/PD simulations, machine learning, and mechanistic mathematic models to pave the way towards personalized medicine. Leveraging different kinds of data through novel quantitative approaches will guide practitioners in decision making in the following decades.

Esteban A. Hernandez-Vargas
Juriquilla, Mexico
January 2022

Acknowledgments

I feel truly fortunate to have had the opportunity to work with all the authors that elaborated fantastic chapters for this book, to all of them, thanks.

I acknowledge funding institutions that supported my research work such as the Universidad Nacional Autonoma de Mexico (UNAM) - PAPITT DGAPA with number IA102521 and the Mexican National Council of Science and Technology (CONACYT).

Finally, I thank the Elsevier team composed of Sonnini Ruiz Yura, Emily Joy Grace Thomson, and Prem Kumar Kaliamoorthi for the great support and guidance throughout the publishing process. I am also grateful to all the anonymous referees for carefully reviewing and selecting the final chapters of this book.

Esteban A. Hernandez-Vargas
Juriquilla, Mexico
January 2022

CHAPTER

1

Closing the loop in personalized medicine

Esteban A. Hernandez-Vargas

Institute of Mathematics, UNAM, Juriquilla, Mexico

Technical aspects that are critical for the development of effective therapeutic and prophylactic approaches to tackle human diseases remain largely fragmented. Scheduling and tailoring drug doses have been performed mainly based on integrating clinical observations to treat patients. However, this empirical approach is infeasible by the steep increase in the number of different pieces of information and the complexity of large datasets. Therefore, a systematic and tractable approach that integrates a variety of biological and medical research data is crucial to harness knowledge towards personalized medicine.

Digital twins are mathematical models and computational algorithms trained with clinical data that allow for the simulation of a virtual patient computationally to develop and test therapies more quickly and economically (Fig. 1.1). This is central to dissect new underlying mechanisms that can lead to better diagnosis, prognosis, and treatment. The value of a model-based approach to drug development for decision making at preclinical development phases has been largely advocated by pharmaceutical companies [1].

Mathematical modeling is an abstraction of a system based on mathematical terms to study the effects of different components and consequently make predictions. Mathematical modeling approaches have been playing a central role in the description of different applications in engineering and natural and social sciences [2]. The major benefit of mathematical modeling is that it can lead to testable experimental predictions [3]. In fact, mathematical modeling is among the renowned methods used to study several aspects of viral infection diseases from virus–host interactions to complex immune response systems and therapies [4].

Feedback Control for Personalized Medicine
https://doi.org/10.1016/B978-0-32-390171-0.00010-X

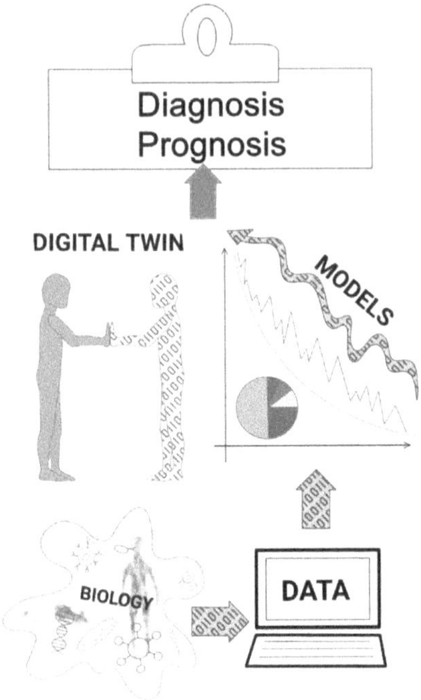

FIGURE 1.1 **Towards personalized medicine with digital twins.** Mathematical modeling and computational simulations are key elements to tailor drugs against diseases. Thus, from datasets the construction of models is fundamental to evaluate in silico the possible outcomes of drugs at the host level.

Personalized medicine is advanced in oncology and HIV where targeted therapies have been available for nearly two decades. However, the development of digital medicine in personalized medicine is still largely fragmented [5], [6]. For example, the Swedish Digital Twin Consortium in [5] proposed a strategy based on (i) constructing digital twins to treat them with thousands of drugs to identify the best-performing drug and, subsequently, (ii) treating the patient with this drug.

Different disciplines can contribute to this emerging field with another perspective. *Control theory* can influence the behavior of a system by modifying its inputs with a feedback strategy to achieve a goal. In this context, control theory can make use of a digital replica of a patient to tailor and schedule therapies for a given disease (see Fig. 1.2). There are several examples of that in the literature, ranging from infectious diseases [4], [7–14] and artificial pancreas [9], [15–17] to cancer [18], [19] and other biological applications [20–22].

Nowadays, pharmaceutical companies have taken a strategic initiative to promote the use of modeling approaches within drug projects [1]. Drug administration can be classically divided into two major phases: a *pharma-*

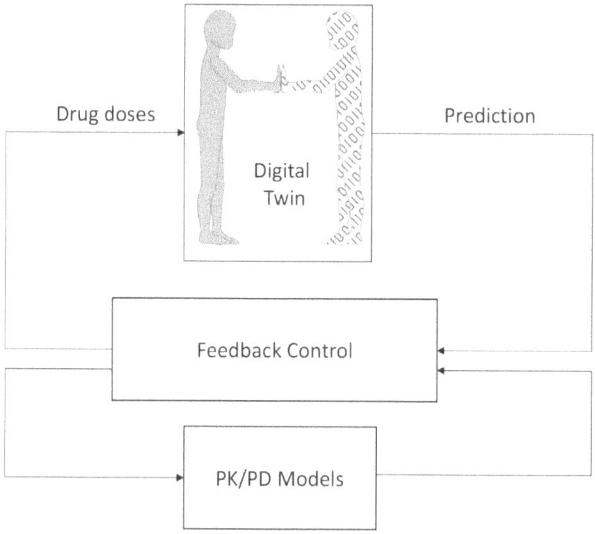

FIGURE 1.2 **Closing the loop with digital twins.** The feedback control will schedule and optimize drug doses based on predictions of a digital twin and using PK/PD models of the drugs to be used.

cokinetic (PK) phase that relates dose and frequency to drug level–time in the body and a *pharmacodynamic* (PD) phase that relates the concentration of the drug to the magnitude of the effects produced [23]. PK/PD modeling and simulation can provide a framework to design feedback strategies and predict the effect time courses resulting from the treatment based on interactions between drugs on a digital twin (see Fig. 1.2). Furthermore, PK/PD dynamics together with mechanistic mathematical models have risen as a promising tool to quantify infectious diseases, facilitating the identification of potential therapeutic targets and vaccine improvement [24–26].

This book presents a set of additional 10 chapters introducing how control theoretical approaches can help to tailor therapies for different diseases, in particular, infectious diseases. Pathogens are a major threat to humankind with about 16 million deaths annually worldwide. Acute infections are caused by pathogens with high growth rates promoting an infection that is cleared in a short period, for example, influenza or COVID-19. On the other hand, chronic infections are caused by pathogens with low growth rates promoting an infection that is persistent for long periods, for example, HIV and Hepatitis.

Antimicrobials such as antibiotics and antivirals are powerful weapons to fight against pathogenic microorganisms. However, the misuse and overuse of antimicrobials have led to drug-resistant strains. In this respect, the World Health Organization (WHO) has reported that antimicrobial resistance (AMR) is a global health problem [27]. This has been exposed by a growing list of viral infections; for instance, HIV resistance has been reported with a range of 7–40% in patients under antiretroviral therapies [28]. Therefore, personalized medicine is paramount to schedule optimal drug doses to optimally treat patients and reduce AMR.

Most of the chapters in the book dealing with acute and chronic viral infections consider variations of *the target cell limited model* (see Fig. 1.3). This is the most used and accepted mathematical model to represent acute viral infection and chronic viral infection dynamics, including HIV [29–31], Hepatitis [32], influenza [33–37], Ebola [26], Dengue [38], and Zika [39]. The target cell limited model is composed of target cells (U) that are replenished with rate S_u and die with rate d_u. A virus (V) infects target cells (U) with rate β. Infected cells are cleared with rate δ. Once cells are productively infected (I), they release the virus at rate p and virus particles are cleared with rate c.

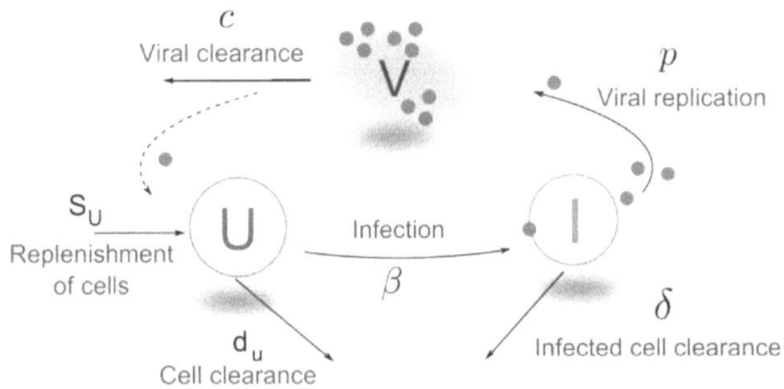

FIGURE 1.3 Target cell model. Panel (a) portrays the scheme of the target cell model which is composed of target cells (U), infected cells (I), and viruses (V). Cells can be replenished with rate S_u, die with rate d_u, and are infected with rate β. Infected cells are cleared with rate δ. Infected cells release the virus at rate p and virus particles are cleared with rate c. This scheme is taken from [34].

Considering the current COVID-19 pandemic, Chapter 2 presents a dynamical characterization of a mathematical model fitted with viral load data of a COVID-19 patient, and subsequently, the optimal fixed-dose antiviral schedule is computed. The control objective for the closed loop in Chapter 2 consists of both maximizing the final value of susceptible/uninfected cells at the end of the infection and minimizing the virus peak.

Monte Carlo simulation is a computational technique to explore and understand the impact of uncertainty in prediction and forecasting models. Chapter 3 reports analytical and individual-based methods for antibiotic dose selection that are based on tools from control theory. Thus, the dosing of antibiotics in this chapter is investigated via *Monte Carlo* simulations from a mathematical PK model. Consequently, the "worst-case" scenario of a patient is employed to evaluate extremely unfavorable conditions of drugs.

A step forward in personalized medicine by control theory can be found in the artificial pancreas for type 1 diabetes mellitus (T1DM) treatment, which is a continuous glucose monitor (CGM) together with a continuous insulin infusion pump and a control strategy to calculate the appropriate dose according to glucose measurements ([40], [41]). Chapter 4 introduces the problem of providing safe glycemic control in T1DM by considering parameter variations, and an offset-free model predictive control (OF-MPC) approach for pulsatile systems is presented. MPCs have a promising future in personalized medicine because of their potential to handle system constraints more easily [40], [42–44]. This type of control strategy solves a finite-horizon optimization problem at every sampling time to obtain an optimal profile of inputs that minimizes an objective function, constrained to a model describing the process evolution.

While Chapter 4 considers a mathematical model based on ordinary equations to make predictions, Chapter 5 merges the power of *deep learning* for glucose prediction. Deep neural networks (DNNs) are an architecture of artificial neural networks which are called "deep" because they are constructed by several layers of nonlinear operations, each one next to the other, in such a way that they can abstract representations of the real world with their weights. Chapter 5 discusses how an automatic control scheme with the integration of several prediction steps based on neural networks could lead to personalized medicine rather than a general population.

Drugs when taken in a pill form cannot provide a continuous action, but it can be assumed that they are instantaneously incorporated into the system at a certain time instant – the drug intake. In this sense, impulsive control arises from the necessity to instantaneously change the state variables of a system. Chapter 6 hypothesizes schemes of dose tailoring towards personalized treatment during influenza infection, where the dose is dynamically adapted according to the viral load evolution. To this end, *inverse optimal control*, which is a stabilizing feedback control law that satisfies a corresponding cost function, is adapted to an impulsive form. Chapter 6 closes by highlighting the implications of the control-based schemes not only to tackle influenza infections but also to combat similar acute infectious diseases such as COVID-19.

A major problem to tailor therapies for different diseases is the lack, quality, or frequency of measurements of different clinical markers. Monitoring these markers in patients can be expensive and for some markers, monitoring is not feasible. An alternative solution to this problem is the computational estimation of the clinical marker. Several estimator schemes require previous knowledge of the model and parameters, but this can be difficult for many medical applications. Among different ways to make an estimator, *filters* are very common in control theory to reduce noise or to estimate variables. For example, the most popular is the *Kalman filter*, which can provide estimates of unknown variables given the measurements observed over time. Chapter 7 presents a continuous-time filter for a class of stochastic nonlinear polynomial systems driven by additive white Gaussian noises. To show the potential use of filters in infectious diseases, this approach and the extended Kalman filter are applied for two different models widely used for modeling viral infectious diseases.

Quantitative methods to evaluate feedback control mechanisms applied by the immune system to control HIV infection are addressed in Chapter 8. While combined antiretroviral treatment (cART) cannot eradicate the virus, it can potently suppress the virus to undetectable levels in the blood, improve immunologic parameters, minimize the emergence of resistance, and protect from opportunistic infections. Although HIV can nowadays be considered as a complex chronic disease, current reports caution that the ongoing costs of HIV therapy and the increase in age-related illness means that the aggregate costs of HIV care will continue to increase [45]. A very debatable clinical aspect in HIV is *elite control*, which occurs when the immune response suppresses the viral load to undetectable levels without treatment. Applying the reachability paradigm from sliding mode control theory, Chapter 8 assesses the robust performance of elite control of HIV. A *sliding mode* is a control strategy that modifies the dynamics by applying a discontinuous control signal that forces the system to "slide" along a cross-section of the system's desired behavior. This sliding mode reachability condition defines a fundamental immunological criterion to determine the expected increase or decrease in viral load.

Hepatitis C virus (HCV) infection is another chronic viral infection causing major public health problems. HCV infection is asymptomatic, and in about 70% of infected individuals, the virus persists and progresses to chronic liver diseases. Chapter 9 introduces a stochastic model of the molecular machinery involved in cells infected with HCV and the respective immune responses. The scale explored in this chapter is at the intracellular level; the information for biomolecules is encoded in the genetic material called nucleic acids: DNA (deoxyribonucleic acid) and RNA (ribonucleic acid). The technique used here is the *chemical master equation* (CME), which can represent the time evolution for the probability function of the system state. Results in this chapter highlight which proteins play an important role in hepatitis dynamics.

Another important control theoretical application in this book is *oncolytic virus therapy* (OVT), which is an experimental therapy that aims to treat cancer using engineered viruses. By viral replication, oncolytic viruses infect and kill tumor cells by lysis or by triggering immune responses against tumor cells [46]. Chapter 10 presents a nonlinear predictive control theory to tune viral doses. The proposed impulsive MPC is designed based on a mathematical model of oncolytic adenovirus therapy.

This determines the optimal viral load to reduce the number of tumor cells to a therapeutic target within a sensible time.

The last chapter of the book is dedicated to a very important organ of the circulatory system, *the heart*. This specialized muscle contracts regularly and continuously, pumping blood to the body and the lungs. Each heartbeat results from electrical activity originating in a specialized pacemaker region called the sinoatrial node in the healthy heart. Chapter 11 aims to investigate whether the cardiac contractile function can be considered as a relaxation oscillator. This is attributed to a differential equation that has an oscillating solution. Thus, this chapter is dedicated to analyzing whether the isolated heart behaves like a relaxation oscillator.

Following chapters provide an effort to couple control theoretical approaches, PK/PD simulations, machine learning, and mechanistic mathematic models, paving the way towards personalized medicine. Leveraging different kinds of data through novel quantitative approaches will guide practitioners in decision making in the following decades.

References

[1] S.A.G. Visser, et al., Model-based drug discovery: implementation and impact, Drug Discovery Today 18 (15–16) (2013) 764–775.

[2] N.A. Gershenfeld, The Nature of Mathematical Modeling, Cambridge University Press, 1999.

[3] C.J. Tomlin, J.D. Axelrod, Biology by numbers: mathematical modelling in developmental biology, Nature Reviews. Genetics 8 (5) (May 2007) 331–340.

[4] E.A. Hernandez-Vargas, Modeling and Control of Infectious Diseases: With MATLAB and R, 1st edition, Elsevier Academic Press, London, 2019.

[5] B. Björnsson, et al., Digital twins to personalize medicine, Genome Medicine 12 (1) (Dec. 2019) 1–4, https://doi.org/10.1186/S13073-019-0701-3.

[6] Reinhard Laubenbacher, James P. Sluka, James A. Glazier, Using digital twins in viral infection, Science 371 (Apr. 2021) 1105–1106, https://doi.org/10.1002/cpt.2160.

[7] P. di Giamberardino, D. Iacoviello, HIV infection control: a constructive algorithm for a state-based switching control, International Journal of Control, Automation, and Systems 16 (3) (Jun. 2018) 1469–1473, https://doi.org/10.1007/s12555-017-0211-2.

[8] E.A. Hernandez-Vargas, Modeling kick-kill strategies toward HIV cure, Frontiers in Immunology 8 (AUG) (2017), https://doi.org/10.3389/fimmu.2017.00995.

[9] P.S. Rivadeneira, M. Caicedo, A. Ferramosca, A.H. Gonzalez, Impulsive zone model predictive control (iZMPC) for therapeutic treatments: application to HIV dynamics, in: 56th IEEE Conference on Decision and Control, 2017, pp. 1–6.

[10] R. Zurakowski, Nonlinear observer output-feedback MPC treatment scheduling for HIV, Biomedical Engineering Online 10 (40) (2011) 1–16, https://doi.org/10.1186/1475-925X-10-40.

[11] R. Zurakowski, A.R. Teel, A model predictive control based scheduling method for HIV therapy, Journal of Theoretical Biology 238 (2) (Jan. 2006) 368–382, https://doi.org/10.1016/j.jtbi.2005.05.004.

[12] H. Chang, A. Astolfi, Control of HIV infection dynamics, IEEE Control Systems Magazine 28 (2) (2008) 28–39.

[13] H. Chang, A. Astolfi, Control of the HIV infection dynamics with a reduced second order model, IEEE American Control Conference 2 (C) (2007) 4083–4088.

[14] P. Abuin, A. Anderson, A. Ferramosca, E.A. Hernandez-Vargas, A.H. Gonzalez, Characterization of SARS-CoV-2 dynamics in the host, Annual Reviews in Control (2020), https://doi.org/10.1016/j.arcontrol.2020.09.008.
[15] S.M. Lynch, B.W. Bequette, Model predictive control of blood glucose in type I diabetics using subcutaneous glucose measurements, in: Proceedings of the American Control Conference, Vol. 5, 2002, pp. 4039–4043.
[16] Z. Cao, E. Dassau, R. Gondhalekar, F.J. Doyle III, Extremum Seeking Control Based Zone Adaptation for Zone Model Predictive Control in Type 1 Diabetes, 2017, pp. 15639–15644.
[17] P.S. Rivadeneira, J.E. Sereno, Blood Glycemia Reconstruction from Discrete Measurements using an Impulsive Observer, 2017, pp. 15288–15293.
[18] H. Namazi, V.V. Kulish, A. Wong, Mathematical modelling and prediction of the effect of chemotherapy on cancer cells, Scientific Reports 5 (Jan. 2015) 13583, https://doi.org/10.1038/srep13583.
[19] P.M. Altrock, L.L. Liu, F. Michor, The mathematics of cancer: integrating quantitative models, Nature Reviews. Cancer 15 (12) (Nov. 2015) 730–745.
[20] E.v. Nikolaev, E.D. Sontag, Quorum-sensing synchronization of synthetic toggle switches: a design based on monotone dynamical systems theory, PLoS Computational Biology 12 (4) (Apr. 2016) e1004881, https://doi.org/10.1371/journal.pcbi.1004881.
[21] A.-L. Barabasi, N. Gulbahce, J. Loscalzo, Network medicine: a network-based approach to human disease, Nature Reviews. Genetics 12 (1) (2011) 56–68, https://doi.org/10.1038/nrg2918.
[22] C. Nowzari, V. Preciado, G. Pappas, Optimal resource allocation for control of networked epidemic models, IEEE Transactions on Control of Network Systems 5870 (c) (2015) 1, https://doi.org/10.1109/TCNS.2015.2482221.
[23] B. Meibohm, H. Derendorf, Basic concepts of PK/PD modelling, International Journal of Clinical Pharmacology and Therapeutics 35 (10) (1997) 401–413.
[24] A. Boianelli, N. Sharma-Chawla, D. Bruder, E.A. Hernandez-Vargas, Oseltamivir PK/PD modeling and simulation to evaluate treatment strategies against influenza-pneumococcus coinfection, Frontiers in Cellular and Infection Microbiology 6 (2016).
[25] C. Parra-Rojas, V.K. Nguyen, G. Hernandez-Mejia, E.A. Hernandez-Vargas, Neuraminidase inhibitors in influenza treatment and prevention–is it time to call it a day?, Viruses 10 (9) (Aug. 2018) 454, https://doi.org/10.3390/v10090454.
[26] V.K. Nguyen, E.A. Hernandez-Vargas, Windows of opportunity for Ebola virus infection treatment and vaccination, Scientific Reports 7 (1) (2017) 8975.
[27] WHO, WHO | Antimicrobial Resistance, World Health Organization, 2016.
[28] WHO, WHO Fact sheet; Global HIV situation, http://www.who.int/gho/hiv/en/, 2019.
[29] A.S. Perelson, Modelling viral and immune system dynamics, Nature Reviews. Immunology 2 (1) (2002) 28–36.
[30] J.M. Conway, D. Coombs, A stochastic model of latently infected cell reactivation and viral blip generation in treated HIV patients, PLoS Computational Biology 7 (4) (2011), https://doi.org/10.1371/journal.pcbi.1002033.
[31] J.M. Conway, A.S. Perelson, Post-treatment control of HIV infection, Proceedings of the National Academy of Sciences 6 (1) (2015) 201419162.
[32] A.S. Perelson, J. Guedj, Modelling hepatitis C therapy: predicting effects of treatment, Nature Reviews. Gastroenterology and Hepatology 12 (8) (Aug. 2015) 437–445.
[33] C. Beauchemin, A. Handel, A review of mathematical models of influenza A infections within a host or cell culture: lessons learned and challenges ahead, BMC Public Health 11 (Suppl 1) (2011) S7.
[34] A. Boianelli, et al., Modeling influenza virus infection: a roadmap for influenza research, Viruses 7 (10) (2015) 5274–5304.

[35] A. Handel, I.M. Longini, R. Antia, Neuraminidase inhibitor resistance in influenza: assessing the danger of its generation and spread, PLoS Computational Biology 3 (12) (2007) 2456–2464.
[36] A. Handel, I.M. Longini, R. Antia, Towards a quantitative understanding of the within-host dynamics of influenza A infections, Journal of the Royal Society Interface 7 (42) (2010) 35–47.
[37] A.M. Smith, Host-pathogen kinetics during influenza infection and coinfection: insights from predictive modeling, Immunological Reviews 285 (1) (Sep. 2018) 97–112, https://doi.org/10.1111/imr.12692.
[38] N. Nuraini, H. Tasman, E. Soewono, K.A. Sidarto, A with-in host Dengue infection model with immune response, Mathematical and Computer Modelling 49 (5–6) (2009) 1148–1155.
[39] C.E. Osuna, et al., Zika viral dynamics and shedding in rhesus and cynomolgus macaques, Nature Medicine 22 (12) (2016) 1448–1455.
[40] A. Haidar, The artificial pancreas: how closed-loop control is revolutionizing diabetes, IEEE Control Systems 36 (5) (2016) 28–47, https://doi.org/10.1109/MCS.2016.2584318.
[41] R. Gondhalekar, E. Dassau, F.J. Doyle, Periodic zone-MPC with asymmetric costs for outpatient-ready safety of an artificial pancreas to treat type 1 diabetes, Automatica 71 (2016) 237–246, https://doi.org/10.1016/j.automatica.2016.04.015.
[42] P. Sopasakis, P. Patrinos, H. Sarimveis, A. Bemporad, Model predictive control for linear impulsive systems, IEEE Transactions on Automatic Control 60 (8) (2015) 2277–2282, https://doi.org/10.1109/TAC.2014.2380672.
[43] A. El Fathi, V. Gingras, B. Boulet, The artificial pancreas and meal control, IEEE Control Systems (February) (2018).
[44] M.F. Villa-Tamayo, A.J.N. Anelone, P.S. Rivadeneira, Tumor reduction using oncolytic viruses under an impulsive nonlinear estimation and predictive control scheme, IEEE Control Systems Letters 5 (5) (2021) 1705–1710, https://doi.org/10.1109/LCSYS.2020.3043185.
[45] F. Barré-Sinoussi, A.L. Ross, J.-F. Delfraissy, Past, present and future: 30 years of HIV research, Nature Reviews. Microbiology 11 (12) (2013) 877–883.
[46] K. Twumasi-Boateng, J.L. Pettigrew, Y.Y.E. Kwok, J.C. Bell, B.H. Nelson, Oncolytic viruses as engineering platforms for combination immunotherapy, Nature Reviews. Cancer 18 (7) (Apr. 2018) 419–432, https://doi.org/10.1038/s41568-018-0009-4.

CHAPTER

2

Optimal control strategies to tailor antivirals for acute infectious diseases in the host: a study case of COVID-19

Mara Pérez[a], *Pablo Abuin*[a], *Marcelo Actis*[b], *Antonio Ferramosca*[c], *Esteban A. Hernandez-Vargas*[d], *and Alejandro H. González*[a]

[a]Institute of Technological Development for the Chemical Industry (INTEC), CONICET-UNL, Santa Fe, Argentina, [b]Facultad de Ingeniería Química (FIQ), Universidad Nacional del Litoral (UNL) and Consejo Nacional de Investigaciones científicas y técnicas (CONICET), Santa Fe, Argentina, [c]Department of Management, Information and Production Engineering, University of Bergamo, Dalmine, BG, Italy, [d]Institute of Mathematics, UNAM, Juriquilla, Mexico

2.1 Introduction

Mathematical models of within-host infections can be used to characterize pathogen dynamics, optimize drug delivery, uncover biological parameters (including pathogen and infected cell half-lives), and design clinical trials, among others. They have been employed to study chronic (i.e., HIV [1–3], hepatitis B [4,5], hepatitis C [6,7]) and acute (i.e., influenza [8–10], Dengue [11,12], Ebola [13]) infections. Currently, they are based on ordinary differential equations (ODEs), which allows to analyze these systems employing mathematical and computational tools. This way, in-host basic

Feedback Control for Personalized Medicine
https://doi.org/10.1016/B978-0-32-390171-0.00011-1

reproduction numbers ($\mathcal{R}$), the stability of equilibrium states, and analytical/numerical solutions can be computed [14–17]. Most of them are based on the target cell limited model to represent chronic/acute infections according to the infection resolution with respect to the target cell production and natural death rates [18]. This way, the equilibrium states differ from isolated equilibrium points (i.e., disease-free and infected equilibria) for the former to a continuous of equilibrium points (i.e., disease-free equilibrium set) for the latter ones. Note that for acute infections, the only feasible equilibrium is the disease-free one, since the pathogen particles at the end of infection will be cleared independently of the in-host reproduction number [18,8,19]. The existence of a healthy equilibrium set implies that stability analysis can be performed considering equilibrium sets as a generalization of equilibrium points, which gives an environment to employ set theoretic methods [20,21], widely used in the design of set-based controllers, although not fully employed for modeling characterization and control of acute infections. Some preliminary results, which will be discussed later in this chapter, can be found in [22].

The control of infection can be modeled considering immune response mechanisms, where the infection is self-controlled by a combination of nonspecific and specific reactions [23–25] or by drug therapies. The inclusion of pharmacokinetic (PK) and pharmacodynamic (PD) models of drug therapies allows the inclusion of therapeutic effects on pathogen evolution [7,18]. Therefore, the model parameters can be changed exogenously by dose frequency and quantity, naturally limited by the inhibitory potential of the drug (expressed in terms of the drug concentration inhibiting 50% of antigen particles [EC_{50}]) and its cytotoxic effect (expressed in terms of the drug concentration which causes death to 50% of susceptible cells [IC_{50}]) [26]. Moreover, since drugs are normally administered orally or by intravenous injections, instantaneous jumps are observed in the concentration of the drug in some tissues. This is mathematically conceptualized as a discontinuity of the first kind and gives rise to the so-called impulsive control systems [27]. This model representation has been used for optimal control and state-feedback control with constraints for infectious diseases, such as influenza [10,28] and HIV [29,27]. Even though optimal dosages can be computed for chronic and acute models, the unstable healthy equilibrium of the former (under certain circumstances; for details, see [18,15]) and the availability of target cells above a critical level for the latter (as discussed later) involve the duration of drug therapy, with the presence of viral rebounds when therapy is disrupted. This effect has been noted for chronic [30] and acute [31] infections. Taking into account this scenario, in this work, we formalize the existence of an optimal single interval drug delivery such that viral rebounds are avoided. Even though the presented analysis is valid for the target cell limited model for acute infections, taking into account the current worldwide COVID-19 pandemic, we prove our results using an identified model of patients infected with the SARS-CoV-2 virus [32,22,33].

After the introduction given in Section 2.1 the chapter is organized as follows. Section 2.2 presents the general "in-host" models used to represent infectious diseases. Section 2.4 studies the way the antivirals affect the dynamics of the model, emphasizing the fact that the stability analysis conducted in Section 2.3 remains unmodified and, hence, any control strategy must be designed accounting for these details. In Section 2.5 control design able to exploit the stability model characterization is introduced, and its benefits are shown by simulating several cases in Section 2.6. Finally, conclusions are given in Section 2.7.

2.1.1 Notation

First let us introduce some basic notation. We consider $\mathbb{R}^n$ as an n-dimensional Euclidean space equipped with the Euclidean distance between two points defined by $d(x,y) := \|x-y\| = [(x-y)'(x-y)]^{1/2}$. The Euclidean distance from a point x to a set $\mathcal{Y}$ is given by $d(x,\mathcal{Y}) := \|x\|_{\mathcal{Y}} = \inf\{\|x-y\| : y \in \mathcal{Y}\}$.

With $\mathcal{X}$ we will denote the constraint set of $\mathbb{R}^3$, given by

$$\mathcal{X} := \mathbb{R}^3_{\geq 0} = \{(x_1, x_2, x_3) \in \mathbb{R}^3 : x_1 \geq 0, x_2 \geq 0 \text{ and } x_3 \geq 0\}.$$

We will consider $\mathcal{X}$ endowed with the *inherit topology* of $\mathbb{R}^3$, i.e., the open sets are intersections of the open set of $\mathbb{R}^3$ with $\mathcal{X}$. Thus, an open ball in $\mathcal{X}$ with center in x and radius $\varepsilon > 0$ is given by $\mathcal{B}_\varepsilon(x) := \{y \in \mathcal{X} : \|x-y\| < \varepsilon\}$ and an ε-neighborhood of set $\mathcal{Y} \subset \mathcal{X}$ is given by $\mathcal{B}_\varepsilon(\mathcal{Y}) := \{x \in \mathcal{X} : \|x\|_{\mathcal{Y}} < \varepsilon\}$. Letting $x \in \mathcal{Y}$, we say that x is an interior point of $\mathcal{Y}$ if there exist $\varepsilon > 0$ such that $\mathcal{B}_\varepsilon(x) \subseteq \mathcal{Y}$. The interior of $\mathcal{Y}$ is the set of all interior points of $\mathcal{Y}$ and it is denoted by $\text{int}(\mathcal{Y})$.

We will denote with $\to$ the limit of a function to a real number, including infinity. In the case of monotone convergence, decreasing or increasing, we will use the symbols $\searrow$ and $\nearrow$, respectively.

2.2 Review of the target cell limited model for in-host infection

Mathematical models of in-host virus dynamics have shown to be useful to understand the interactions that govern infections and, more importantly, to allow external intervention to moderate their effects [24]. According to recent research in the area [18,22], the following ODEs are used in this work to describe the interaction between uninfected target or susceptible cells U [cells/mm^3], infected cells I [cell/mm^3], and virus V [copies/mL]:

$$\dot{U}(t) = -\beta U(t) V(t), \qquad U(0) = U_0, \tag{2.1a}$$

$$\dot{I}(t) = \beta U(t)V(t) - \delta I(t), \qquad I(0) = I_0, \tag{2.1b}$$

$$\dot{V}(t) = pI(t) - cV(t), \qquad V(0) = V_0, \tag{2.1c}$$

where β [mL.day^{-1}/copies] is the infection rate of healthy U cells by external virus V, δ [day^{-1}] is the death rates of I, p [(copies.mm^3/cell.mL).day^{-1}] is the production rate of free virus from infected cells I, and c [day^{-1}] is the degradation (or clearance) rate of virus V by the immune system.

System (2.1) is positive, which means that $U(t) \geq 0$, $I(t) \geq 0$, and $V(t) \geq 0$, for all $t \geq 0$. We denote by $x(t) := (U(t), I(t), V(t))$ the state vector and by $\mathcal{X} = \mathbb{R}^3_{\geq 0}$ the state constraint set.

The initial conditions of (2.1), which represent a healthy steady state before the infection, are assumed to be $V(t) = 0$, $I(t) = 0$, and $U(t) = U_0 > 0$, for $t < 0$. Then, at time $t = 0$, a small quantity of virions enters the host body and, hence, a discontinuity occurs in $V(t)$. Indeed, $V(t)$ jumps from 0 to a small positive value V_0 at $t = 0$ (formally, $V(t)$ has a discontinuity of the first kind at t_0, i.e., $\lim_{t \to 0^-} V(t) = 0$ while $\lim_{t \to 0^+} V(t) = V_0 > 0$).

Although the solution of (2.1) for $t \geq t_0$, with t_0 being an arbitrary time, is unknown, we know that it depends on the *basic reproduction number* $\mathcal{R} := \frac{\beta p}{c\delta}$ and the initial conditions $(U(t_0), I(t_0), V(t_0)) \in \mathcal{X}$. Since $U(t) \geq 0$, $V(t) \geq 0$, for all $t \geq t_0$, $U(t)$ is a nonincreasing function of t (by 2.1.a). From [22] and [17], if $c \gg \delta$ (as is always the case), system (2.1) can be approximated by $\dot{U}(t) \approx -\beta U(t)V(t)$, $\dot{V}(t) \approx (\frac{\beta p}{c} U(t) - \delta)V(t)$, $I(t) \approx \frac{c}{p} V(t)$. Then, since $U(t_0) > 0$, conditions for V to increase or decrease at t_0 are given by $U(t_0)\mathcal{R} > 1$ and $U(t_0)\mathcal{R} < 1$, respectively. This means that for $U(t_0)\mathcal{R} \leq 1$, $V(t)$ goes asymptotically to zero for $t \to \infty$. Otherwise, if $U(t_0)\mathcal{R} > 1$, $V(t)$ reaches a maximum $\hat{V}$ and then goes asymptotically to zero, for $t \to \infty$. In this latter case it is said that the virus *spreads in the host*, since there is at least one time instant for which $\dot{V} > 0$ [22]. The so-called *critical value of U*, U^*, is defined as

$$U^* := 1/\mathcal{R}, \tag{2.2}$$

where $\mathcal{R}$ is assumed to remain constant for all $t \geq t_0$. The critical value U^* can be seen as the counterpart of the "herd immunity" in the epidemiological SIR-type models, i.e., $U(t)$ reaches U^* approximately at the same time as $V(t)$ and $I(t)$ reach their peaks or, in other words, $V(t)$ and $I(t)$ cannot increase anymore once $U(t)$ is below U^*. This way, conditions $U(t_0)\mathcal{R} > 1$ and $U(t_0)\mathcal{R} \leq 1$ that determine if $V(t)$ increases or decreases for $t \geq t_0$ can be rewritten as $U(t_0) > U^*$ and $U(t_0) \leq U^*$, respectively. In what follows, we assume that $U(0) > U^*$ (or $U(0)\mathcal{R} > 1$), which corresponds to the case of the outbreak of the infection (i.e., the virus does spread in the host), at time $t = 0$.

Let us now define $U_\infty := \lim_{t\to\infty} U(t)$, $V_\infty := \lim_{t\to\infty} V(t)$, and $I_\infty := \lim_{t\to\infty} I(t)$, which are values that depend on $\mathcal{R}$ and the initial conditions $U(t_0)$, $V(t_0)$, $I(t_0)$. According to [22], $V_\infty = I_\infty = 0$, while U_∞ is a value in $(0, U(t_0))$, which will be characterized in the next section.

2.3 Equilibrium characterization and stability

To find the equilibrium set of model (2.1), with initial conditions $(U(t_0), V(t_0), I(t_0)) \in \mathcal{X}$ at an arbitrary time $t_0 \geq 0$, $\dot{U}(t)$, $\dot{I}(t)$, and $\dot{V}(t)$ need to be equaled to zero in (2.1). According to [22,17] there is only one equilibrium set in $\mathcal{X}$, which is a healthy one, and it is defined by

$$\mathcal{X}_s := \{(U, I, V) \in \mathcal{X} : \ I = 0, \ V = 0\}. \tag{2.3}$$

To examine the stability of the equilibrium points in $\mathcal{X}_s$, a first attempt consists in linearizing system (2.1) at some state $x_s := (U_s, I_s, V_s) \in \mathcal{X}_s$ and analyzing the eigenvalues of the Jacobian matrix. As shown in [22], this matrix has one eigenvalue at zero ($\lambda_1 = 0$), one always negative ($\lambda_2 < 0$), and a third one, λ_3, that is negative, zero, or positive depending on if U_s is smaller than, equal to, or greater than U^*, respectively.

Since the maximum eigenvalue λ_3 is the one determining the stability of the system, it is possible to separate set $\mathcal{X}_s$ into two subsets, according to its behavior. Then, a first intuition is that the equilibrium subset

$$\mathcal{X}_s^{st} := \{(U, I, V) \in \mathcal{X} : U \in [0, U^*], \ I = 0, \ V = 0\} \tag{2.4}$$

is stable and that the equilibrium subset

$$\mathcal{X}_s^{un} := \{(U, I, V) \in \mathcal{X} : U \in (U^*, +\infty), \ I = 0, \ V = 0\} \tag{2.5}$$

is unstable. However, this is not a conclusive analysis, given that one of the eigenvalues of the linearized system is null and hence the linear approximation cannot be used to fully determine the stability of a nonlinear system (theorem of Hartman–Grobman [34, Section 2.8, p. 120]).

The formal asymptotic stability (AS) of set $\mathcal{X}_s^{st}$, together with its corresponding domain of attraction (DOA), is analyzed in the next subsection.

2.3.1 Asymptotic stability of the equilibrium sets

A key point to properly analyze the AS of system (2.1) is to consider the stability of the equilibrium sets $\mathcal{X}_s^{st}$ and $\mathcal{X}_s^{un}$, instead of the single points inside them, as defined in Definitions 2.2, 2.3, and 2.5 in the next subsection. Indeed, even when every equilibrium point in $\mathcal{X}_s^{st}$ is ϵ-δ stable, there is no single equilibrium point in such set that is locally attractive.

As stated in Definition 2.5, the AS of $\mathcal{X}_s^{st}$ requires both attractivity and ϵ-δ stability, which are stated in Sections 2.3.3 and 2.3.4, respectively. Finally, in Section 2.3.5 the AS theorem is formally stated.

2.3.2 Stability theory

In this section some basic definitions and results are given concerning the AS of sets and Lyapunov theory in the context of nonlinear continuous-time systems ([20], Appendix B). All the following definitions refer to the system

$$\dot{x}(t) = f(x(t)), \quad x(0) = x_0, \tag{2.6}$$

where x is the system state constrained to be in $\mathcal{X} \subseteq \mathbb{R}^n$, f is a Lipschitz continuous nonlinear function, and $\phi(t; x)$ is the solution for time t and initial condition x.

Definition 2.1 (Equilibrium set). Consider system (2.6) constrained by $\mathcal{X}$. The set $\mathcal{X}_s \subset \mathcal{X}$ is an equilibrium set if each point $x \in \mathcal{X}_s$ is such that $f(x) = 0$ (this implying that $\phi(t; x) = x$ for all $t \geq 0$).

Definition 2.2 (Attractivity of an equilibrium set). Consider system (2.6) constrained by $\mathcal{X}$ and a set $\mathcal{Y} \subseteq \mathcal{X}$. A closed equilibrium set $\mathcal{X}_s \subset \mathcal{X}$ is attractive in $\mathcal{Y}$ if $\lim_{t\to\infty} \|\phi(t; x)\|_{\mathcal{X}_s} = 0$ for all $x \in \mathcal{Y}$. If $\mathcal{Y}$ is an ε-neighborhood of $\mathcal{X}_s$ for some $\eta > 0$, we say that $\mathcal{X}_s$ is locally attractive.

We define the *domain of attraction* (DOA) of an attractive set $\mathcal{X}_s$ for the system (2.6) to be the set of all initial states x such that $\|\phi(t; x)\|_{\mathcal{X}_s} \to 0$ as $t \to \infty$. We use the term *region of attraction* to denote any set of initial states contained in the DOA.

A closed subset of an attractive set (for instance, a single equilibrium point) is not necessarily attractive. On the other hand, any set containing an attractive set is attractive, so the significant attractivity concept in a constrained system is given by the smallest one. Indeed, given two different attractive sets in $\mathcal{X}$ with the same DOA, one must be contained in the other. So the family of all attractive sets in $\mathcal{X}$ with the same DOA is a totally ordered set under the set inclusion (nested family). An arbitrary (finite, countable, or uncountable) intersection of nested nonempty closed subsets of a compact space is a nonempty compact set [35]. Then if one element of the family is bounded, and therefore compact, the intersection of all the family is a nonempty compact set. This set is the *smallest attractive set* we are referring to.

Definition 2.3 (Local ϵ-δ stability of an equilibrium set). Consider system (2.6) constrained by $\mathcal{X}$. A closed equilibrium set $\mathcal{X}_s \subset \mathcal{X}$ is ϵ-δ locally stable if for all $\epsilon > 0$ there exists $\delta > 0$ such that if $\|x\|_{\mathcal{X}_s} < \delta$, then $\|\phi(t; x)\|_{\mathcal{X}_s} < \epsilon$, for all $t \geq 0$.

Unlike attractive sets, a set containing a locally ϵ-δ stable equilibrium set is not necessarily locally ϵ-δ stable. Moreover, a closed subset of a locally ϵ-δ stable equilibrium set (for instance, a single equilibrium point) is not necessarily locally ϵ-δ stable. However, any (finite) union of equilibrium sets locally ϵ-δ stable is also locally ϵ-δ stable. So the significant stability concept in a constrained system is given by the largest one.

Although a finite union of equilibrium set locally ϵ-δ stable is also locally ϵ-δ stable, in general we cannot extend this result to the case of arbitrary unions of points. Thus, even when every equilibrium point of an equilibrium set is locally ϵ-δ stable, we cannot ensure that the whole set would be locally ϵ-δ stable. This is due to the fact that given a fixed ϵ the δ chosen for each point depends on the point, so the infimum of them could be zero. However, if in addition we also assume that the set is compact, then the stability of the set can be inherited from the stability of its points.

Lemma 2.4. *Let $\mathcal{X}_s$ be a compact equilibrium set. If every $x_s \in \mathcal{X}_s$ is ϵ-δ locally stable, then $\mathcal{X}_s$ is ϵ-δ locally stable.*

Proof. Given $\epsilon > 0$, there exists $\delta = \delta(x_s) > 0$ for each $x_s \in \mathcal{X}_s$ such that if $x \in B_{\delta(x_s)}(x_s)$, then $\phi(t; x) \in B_\epsilon(x_s)$ for $t \geq 0$. The family of δ-balls form an open cover of $\mathcal{X}_s$. Let us denote the union of this cover V, i.e., $V := \bigcup\{B_{\delta(x_s)}(x_s) : x_s \in \mathcal{X}_s\}$. Since $\mathcal{X}_s$ is compact and the complement of V is closed, the distance between them is strictly positive, i.e., $\delta^* := d(\mathcal{X}_s, V^c) > 0$. Therefore, the δ^*-neighborhood of the equilibrium set $\mathcal{X}_s$ is contained in V. Thus if $x \in B_{\delta^*}(\mathcal{X}_s) \subset V$, then $\phi(t; x) \in B_\epsilon(\mathcal{X}_s)$ for $t \geq 0$. Therefore $\mathcal{X}_s$ is ϵ-δ locally stable. □

Definition 2.5 (Asymptotic stability of an equilibrium set). Consider system (2.6) constrained by $\mathcal{X}$ and a set $\mathcal{Y} \subseteq \mathcal{X}$. A closed equilibrium set $\mathcal{X}_s \subset \mathcal{X}$ is AS in $\mathcal{Y}$ if it is ϵ-δ locally stable and attractive in $\mathcal{Y}$.

Next, the theorem of Lyapunov, which refers to single equilibrium points and provides sufficient conditions for both local ϵ-δ stability and AS, is introduced.

Theorem 2.6. (Lyapunov's stability theorem [36, Theorem 4.1]) *Consider system (2.6) constrained by $\mathcal{X}$ and an equilibrium state $x_s \in \mathcal{X}_s$. Let $\mathcal{Y} \subset \mathcal{X}$ be a neighborhood of x_s and consider a function $V(x) : \mathcal{Y} \to \mathbb{R}$ such that $V(x) > 0$ for $x \neq x_s$, $V(x_s) = 0$ and $\dot{V}(x(t)) \leq 0$, denoted as Lyapunov function. Then, the existence of such a function in a neighborhood of x_s implies that $x_s \in \mathcal{X}_s$ is locally ϵ-δ stable in $\mathcal{Y}$. If in addition $\dot{V}(x(t)) < 0$ for all $x \neq x_s$, then x_s is AS in $\mathcal{Y}$.*

2.3.3 Attractivity of set $\mathcal{X}_s^{st}$

Before we deal with the attractivity of set $\mathcal{X}_s^{st}$, we need to delve into the behavior of U_∞. As is presented in [22], U_∞ can be expressed as a function of initial conditions as follows:

$$U_\infty(U(t_0), I(t_0), V(t_0)) = -\frac{W(-\mathcal{R}U(t_0)e^{-\mathcal{R}(U(t_0)+I(t_0)+\frac{\delta}{p}V(t_0))})}{\mathcal{R}}, \qquad (2.7)$$

where $W(\cdot)$ is (the principal branch of) the Lambert function and $(U(t_0), I(t_0), V(t_0))$ are arbitrary initial conditions at a given time $t_0 \geq 0$. For each $\varepsilon \geq 0$ let us define a domain of $\mathcal{X}$ given by

$$\Omega(\varepsilon) = \{(U, I, V) \in \mathcal{X} : I \geq \varepsilon, V \geq \varepsilon\}. \qquad (2.8)$$

The following lemma describes the behavior of the maximum of U_∞ on each $\Omega(\varepsilon)$.

Lemma 2.7 (Maximum of the function U_∞). *Consider the function U_∞ given by (2.7) and for each $\varepsilon \geq 0$ the domains $\Omega(\varepsilon)$ given by (2.8). Then the maximum of $U_\infty(U, I, V)$ in $\Omega(\varepsilon)$ is reached in $(U^*, \varepsilon, \varepsilon)$. In particular, the maximum value of U_∞ over $\Omega(0)$ is reached in $(U^*, 0, 0)$ and is given by $U_\infty(U^*, 0, 0) = U^*$, where $U^* = 1/\mathcal{R}$.*

Proof. According to (2.7), U_∞ can be written as

$$U_\infty(U, I, V) = -\frac{W(-f(U, I, V))}{\mathcal{R}},$$

with $f(U, I, V) = \mathcal{R}Ue^{-\mathcal{R}(U+I+\delta/pV)}$. Since $-W(-\cdot)$ is an increasing (injective) function, $U_\infty(U, I, V)$ achieves its maximum over $\Omega(\varepsilon)$ at the same values as $f(U, I, V)$. Then, we focus our attention on finding the maximum (and the maximizing variables) of $f(U, I, V)$.

Through the change of variables $x = \mathcal{R}U$ and $y = \mathcal{R}(I + \delta/pV)$, f can be studied as a function of the form $g(x, y) = xe^{-(x+y)}$. Note that $(U, I, V) \in \Omega(\varepsilon)$ if and only if $x \geq 0$ and $y \geq \eta$, where $\eta := \mathcal{R}(1 + \frac{\delta}{p})\varepsilon \geq 0$. Therefore, to find extremes of f in $\Omega(\varepsilon)$ it is enough to study the extreme points of g over $\Omega' = \{(x, y) \in \mathbb{R}^2_{\geq 0} : y \geq \eta\}$.

Since $\nabla g = [(1 - x)e^{-(x+y)}, -xe^{-(x+y)}]$ does not vanish and $g \to 0$ when $\|(x, y)\| \to \infty$, the maximum is reached at the boundaries of Ω'. A simple analysis shows that g restricted to the boundary of Ω' achieves its maximum in $(1, \eta)$. This means that $f(U, I, V)$ achieves its maximum in $U = 1/\mathcal{R} = U^*$ and $I = V = \varepsilon$.

In particular, when $\varepsilon = 0$, $f(U, I, V)$ reaches its maximum in $(U^*, 0, 0)$. Furthermore,

$$U_\infty(U^*, 0, 0) = -\frac{W(-f(U^*, 0, 0))}{\mathcal{R}} = -\frac{W(-1/e)}{\mathcal{R}} = \frac{1}{\mathcal{R}} = U^*,$$

which concludes the proof. □

Let us now deal with attractivity of set $\mathcal{X}_s^{st}$. According to Definition 2.2, any set containing an attractive set is also attractive. So, we are in fact interested in finding the smallest closed attractive set in $\mathcal{X}\backslash\mathcal{X}_s^{un}$.

Theorem 2.8 (Attractivity of $\mathcal{X}_s^{st}$). *Consider system* (2.1) *constrained by* $\mathcal{X}$. *The set* $\mathcal{X}_s^{st}$ *defined in* (2.4) *is the smallest attractive set in* $\mathcal{X}\backslash\mathcal{X}_s^{un}$. *Furthermore,* $\mathcal{X}_s^{un}$, *defined in* (2.5), *is not attractive.*

Proof. The proof is divided into two parts. First it is proved that $\mathcal{X}_s^{st}$ is an attractive set and then it is proved that it is the smallest one.

Attractivity of $\mathcal{X}_s^{st}$**:** To prove the attractivity of $\mathcal{X}_s^{st}$ in $\mathcal{X}$ (and to show that $\mathcal{X}_s^{un}$ is not attractive) we need to prove that $U_\infty \in [0, U^*]$ for any initial conditions and values of $\mathcal{R}$. The minimum of U_∞ is given by $U_\infty = 0$, and it is reached when $U(t_0) = 0$ (for any value of $\mathcal{R}$, $I(t_0)$ and $V(t_0)$). The maximum of U_∞, on the other hand, is given by $U_\infty = U^*$, and it is reached only when $U(t_0) = U^*$ and $I(t_0) = V(t_0) = 0$ (for any value of $\mathcal{R}$), as shown in Lemma 2.7. Then, for any $(U(t_0), I(t_0), V(t_0)) \in \mathcal{X}$ and $\mathcal{R} > 0$, $U_\infty(\mathcal{R}, U(t_0), I(t_0), V(t_0)) \in [0, U^*]$, which means that $\mathcal{X}_s^{st}$ is attractive, and the proof of attractivity is complete.

Fig. 2.1 shows how U_∞ behaves as function of $U(t_0)$ and $V(t_0)$, when $I(t_0) = 0$ and $I(t_0) = 5e^5$. The first one is the scenario corresponding to $t = 0$, when a certain amount of virus enters the healthy host.

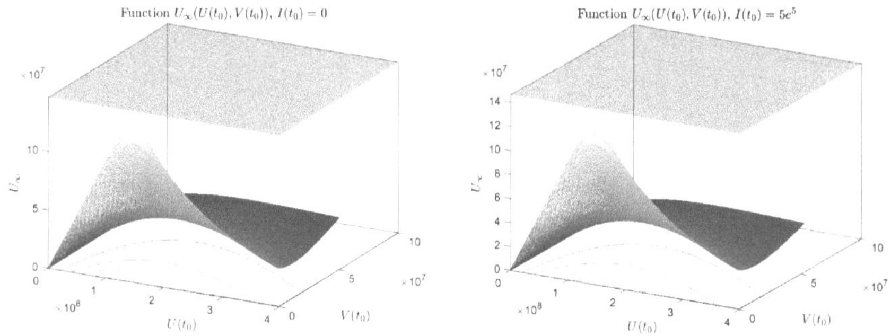

(a) U_∞ as a function of $U(t_0)$ and $V(t_0)$, when $I(t_0) = 0$. The orange (light gray in print version) plane represents $U^* = 1/\mathcal{R} = 1.5e^8$. The maximum of U_∞ is reached when $U(t_0) = U^*$ and $V(t_0) = 0$ and is given by U^*. Patient A.

(b) U_∞ as a function of $U(t_0)$ and $V(t_0)$, when $I(t_0) = 5e^5$. The orange (light gray in print version) plane represents $U^* = 1/\mathcal{R} = 1.5e^8$. The maximum of U_∞ is reached when $U(t_0) = U^*$ and $V(t_0) = 0$ and is smaller than U^*. Patient A.

FIGURE 2.1 Function $U_\infty(U(t_0), V(t_0))$ for different values of $I(t_0)$.

$\mathcal{X}_s^{st}$ ***is the smallest attractive set:*** It is clear from the previous analysis that any initial state $x(t_0) = (U(t_0), I(t_0), V(t_0)) \in \mathcal{X}$ converges to a state $x_\infty = (U_\infty, 0, 0)$ with $U_\infty \in [0, U^*]$. This means that $\mathcal{X}_s^{un}$ is not attractive for any point in $\mathcal{X}\backslash\mathcal{X}_s$. However, to show that $\mathcal{X}_s^{st}$ is the smallest attractive set, we need to prove that every point $x_s \in \mathcal{X}_s^{st}$ is necessary for the attractiveness.

Let us consider an initial state of the form $(U^*, I(t_0), 0)$ with $I(t_0) \geq 0$. Since W is a bijective function from $(-1/e, 0)$ to $(-1, 0)$, $U_\infty(\mathcal{R}, U^*, \cdot, 0)$ is bijective from $(0, +\infty)$ to $(0, U^*)$. Hence for every point $x_s \in \text{int}(\mathcal{X}_s^{st})$ there exists $I(t_0) \geq 0$ such that the initial state $(U^*, I(t_0), 0)$ converges to x_s. Since every interior point of $\mathcal{X}_s^{st}$ is necessary for the attractiveness, the smallest closed attractive set is $\mathcal{X}_s^{st}$, and the proof is concluded. □

2.3.4 Local ϵ-δ stability of $\mathcal{X}_s^{st}$

The next theorem shows the formal Lyapunov (or ϵ-δ) stability of the equilibrium set $\mathcal{X}_s^{st}$.

Theorem 2.9 (Local ϵ-δ stability of $\mathcal{X}_s^{st}$). *Consider system* (2.1) *constrained by* $\mathcal{X}$. *Then, the equilibrium set* $\mathcal{X}_s^{st}$ *defined in* (2.4) *is the largest locally* ϵ-δ *stable one.*

Proof. We proceed by analyzing the stability of single equilibrium points $\bar{x} := (\bar{U}, 0, 0)$, with $\bar{U} \in (0, U_0]$ (i.e., $\bar{x} \in \mathcal{X}_s \backslash \{(0, 0, 0)\}$). For each $\bar{x}$ let us consider the following Lyapunov function candidate:

$$J(x) := U - \bar{U} - \bar{U} \ln\left(\frac{U}{\bar{U}}\right) + I + \frac{\delta}{p} V. \tag{2.9}$$

This function is continuous in $\mathcal{X}$ and is positive for all nonnegative $x \neq \bar{x}$ and $J(\bar{x}) = 0$. Furthermore, for $x(t) \in \mathcal{X}$ and $t \geq 0$ we have

$$\begin{aligned}
\dot{J}(x(t)) = \frac{\partial J}{\partial x}\dot{x}(t) &= \begin{bmatrix} \frac{dJ}{dU} & \frac{dJ}{dI} & \frac{dJ}{dV} \end{bmatrix} \begin{bmatrix} -\beta U(t)V(t) \\ \beta U(t)V(t) - \delta I(t) \\ pI(t) - cV(t) \end{bmatrix} \\
&= \begin{bmatrix} (1 - \frac{\bar{U}}{U(t)}) & 1 & \frac{\delta}{p} \end{bmatrix} \begin{bmatrix} -\beta U(t)V(t) \\ \beta U(t)V(t) - \delta I(t) \\ pI(t) - cV(t) \end{bmatrix} \\
&= (-\beta U(t)V(t) + \bar{U}\beta V(t)) + (\beta U(t)V(t) - \delta I(t)) \\
&\quad + \left(\delta I(t) - \frac{\delta c}{p} V(t)\right) \\
&= \bar{U}\beta V(t) - \frac{\delta c}{p} V(t) = V(t)\left(\bar{U}\beta - \frac{\delta c}{p}\right),
\end{aligned}$$

where $\dot{x}(t)$ represents system (2.1). Function $\dot{J}(x(t))$ depends on $x(t)$ only through $V(t)$. So, independently of the value of the parameter $\bar{U}$, $\dot{J}(x(t)) = 0$ for $V(t) \equiv 0$. This means that for any single $x(0) \in \mathcal{X}_s$, $V(0) = I(0) = 0$ and, hence, $V(t) = 0$, for all $t \geq 0$. So $\dot{J}(x(t))$ is null for any $x(0) \in \mathcal{X}_s$ (i.e., it is not only null for $x(0) = \bar{x}$ but for any $x(0) \in \mathcal{X}_s$).

On the other hand, for $x(0) \notin \mathcal{X}_s$, function $\dot{J}(x(t))$ is negative, zero, or positive, depending on if the parameter $\bar{U}$ is smaller than, equal to, or greater than $U^* = \frac{\delta c}{\beta p}$, respectively, and this holds for all $x(0) \in \mathcal{X}$ and $t \geq 0$. So, for any $\bar{x} \in \mathcal{X}_s^{st}$, $\dot{J}(x(t)) \leq 0$ (particularly, for $\bar{x} = (\bar{U}, 0, 0) = (U^*, 0, 0)$, $\dot{J}(x(t)) = 0$, for all $x(0) \in \mathcal{X}$ and $t \geq 0$), which means that each $\bar{x} \in \mathcal{X}_s^{st}$ is locally ϵ-δ stable (see Theorem 2.6).

Finally, when $\bar{U} = 0$, i.e., $\bar{x} = (0, 0, 0)$, we define the Lyapunov functional as $J(x) = U - I + \delta/pV$ and we proceed analogously as before to prove the local ϵ-δ stability of the origin.

Therefore, since every state in $\mathcal{X}_s^{st}$ is locally ϵ-δ stable and $\mathcal{X}_s^{st}$ is compact, by Lemma 2.4, the whole set $\mathcal{X}_s^{st}$ is locally ϵ-δ stable.

Finally, since $\mathcal{X}_s^{st}$ is attractive in $\mathcal{X} \backslash \mathcal{X}_s^{un}$, it is impossible for any $x \in \mathcal{X}_s^{un}$ to be ϵ-δ stable, which implies that $\mathcal{X}_s^{st}$ is also the largest locally ϵ-δ stable set in $\mathcal{X}_s$, which completes the proof. □

Remark 2.10. In the latter proof, if we pick a particular $\bar{x} \in \mathcal{X}_s^{st}$, then $\dot{J}(x(t))$ is null not only for $x(0) = \bar{x}$, but for all $x(0) \in \mathcal{X}_s^{st}$, since in this case, $V(t) = 0$, for $t \geq 0$. This means that it is not true that $\dot{J}(x(t)) < 0$ for every $x \neq \bar{x}$, and this is the reason why we cannot use the last part of Theorem 2.6 to ensure the AS of particular equilibrium points (or subsets of $\mathcal{X}_s^{st}$). In fact, they are ϵ-δ stable, but not attractive.

2.3.5 Asymptotic stability of $\mathcal{X}_s^{st}$

In the next theorem, based on the previous results concerning the attractivity and ϵ-δ stability of $\mathcal{X}_s^{st}$, the AS is formally stated.

Theorem 2.11. *Consider system* (2.1) *constrained by the positive set $\mathcal{X}$. Then, the set $\mathcal{X}_s^{st}$ defined in* (2.4) *is the unique AS equilibrium set, with a DOA given by $\mathcal{X} \backslash \mathcal{X}_s^{un}$. Furthermore, $\mathcal{X}_s^{un}$ is unstable.*

Proof. The proof follows from Theorem 2.8, which states that $\mathcal{X}_s^{st}$ is the smallest attractive in $\mathcal{X}$, and Theorem 2.9, which states that $\mathcal{X}_s^{st}$ is the largest locally ϵ-δ stable set in $\mathcal{X}$. □

Fig. 2.2 shows phase portrait plots of system (2.1), corresponding to different initial conditions.

2.3.6 U_∞ as a function of initial conditions

In this section some characteristics of system (2.1) concerning the value of U_∞ as a function of the reproduction number $\mathcal{R}$ and the initial conditions are analyzed. Consider the next property.

Property 2.12. *Consider system* (2.1) *with arbitrary initial conditions* $(U(t_0), I(t_0), V(t_0)) \in \mathcal{X}$, *for some* $t_0 \geq 0$. *Then:*

i. For any value of $U(t_0) > 0$, $I(t_0) > 0$, $V(t_0) > 0$, $U_\infty(\mathcal{R}, U(t_0), I(t_0), V(t_0)) \to 0$, when $\mathcal{R} \to \infty$; while $U_\infty(\mathcal{R}, U(t_0), I(t_0), V(t_0))$ remains close to $U(t_0)$ when $\mathcal{R} \to 0$.

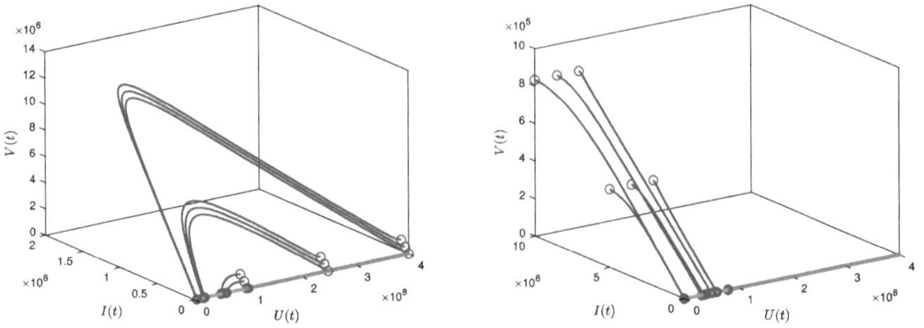

(a) (b)

FIGURE 2.2 Phase portrait for virtual patient A, described in Section 2.3.7. (a) Case $U(t_0) > U^*$. States arbitrarily close to $\mathcal{X}_s^{un}$ (in red (gray in print version)) converge to $\mathcal{X}_s^{st}$ (in green (mid gray in print version)), so the virus spreads in the host. (b) Case $U(t_0) < U^*$. States arbitrarily close to $\mathcal{X}_s^{st}$ converge to $\mathcal{X}_s^{st}$, so the virus does not spread in the host. Empty circles represent the initial state, while solid circles represent final states.

ii. *For $U(t_0) > U^*$ and fixed $I(t_0) > 0$, $V(t_0) > 0$, and $\mathcal{R} > 0$, $U_\infty(\mathcal{R}, U(t_0), I(t_0), V(t_0))$ decreases when $U(t_0)$ increase, and $U_\infty(\mathcal{R}, U(t_0), I(t_0), V(t_0)) < U^*$. This means that the closer $U(t_0)$ is to U^* from above, the closer will U_∞ be to U^* from below.*

iii. *For $U(t_0) < U^*$ and fixed $I(t_0) > 0$, $V(t_0) > 0$, and $\mathcal{R} > 0$, $U_\infty(\mathcal{R}, U(t_0), I(t_0), V(t_0))$ increases with $U(t_0)$, and $U_\infty(\mathcal{R}, U(t_0), I(t_0), V(t_0)) < U^*$. This means that smaller values of $U(t_0)$ produce smaller values of U_∞, both below U^*.*

iv. *For any fixed $U(t_0)$ and $\mathcal{R} > 0$, $U_\infty(\mathcal{R}, U(t_0), I(t_0), V(t_0))$ decrease with $I(t_0)$ and $V(t_0)$, and $U_\infty(\mathcal{R}, U(t_0), I(t_0), V(t_0)) \leq U^*$.*

v. *For fixed $\mathcal{R} > 0$, $U(t_0) = U^*$, and $I(t_0) = V(t_0) = 0$, $U_\infty(\mathcal{R}, U(t_0), I(t_0), V(t_0))$ reaches its maximum over $\mathcal{X}$, and the maximum value is given by U^* (see Lemma 2.7).*

The proof of the properties is omitted for brevity. However, Figs. 2.1 and 2.3 show how U_∞ behaves for different values of initial conditions.

2.3.7 Simulation example

All along this work we use a virtual patient, denoted as patient A, to demonstrate the results of each section. The parameters of patient A were estimated by using viral load data of a COVID-19 patient who was RT-PCR-positive – reported in [32] and used in [22,33] – given by $\beta = 1.35 \times 10^{-7}$, $\delta = 0.61$, $p = 0.2$, and $c = 2.4$. The initial conditions are given by $U_0 = 4 \times 10^8$, $I_0 = 0$, and $V_0 = 0.31$. Furthermore, the reproduction number is $\mathcal{R} = 1.84 \times 10^{-8}$, while the critical value for the susceptible cells is $U^* = 5.44 \times 10^7$. The final value of U (if no antiviral treatment is applied) is given by $U_\infty = 2.57 \times 10^5$, which means that the area under the curve of V, given

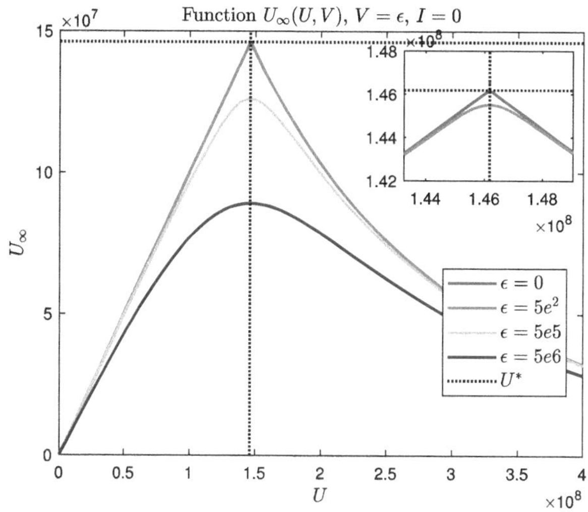

FIGURE 2.3 Function $U_\infty(U, V)$, for different values of ϵ, when $U \in [0, U_{\max}]$, $V = \epsilon$, and $I = 0$. As can be seen, the supremum of U_∞ (given by U^*) is achieved when $U = U^*$ and $I \to 0$.

by $AUC_V := \int_0^\infty V(t)dt$, is 5.45×10^7 (see Remark 2.13). The peak of V is given by $\hat{V} = 1.98 \times 10^7$. Fig. 2.4 shows the time response corresponding to patient A. As predicted, U_∞ is (significantly) smaller than U^*, which means that antivirals reducing (even for a finite period of time) either p or β will increase U_∞ and, hence, will reduce the AUC_V and, probably, the peak of V.

Remark 2.13. Note that from system (2.1), by replacing (2.1a) in (2.1b) and then, after rearranging, in (2.1c), $V(t)$ can be expressed as $V(t) = \frac{1}{c}[-\frac{p}{\delta}(\dot{I}(t) + \dot{U}(t)) - \dot{V}(t)]$. Therefore, the area under the curve of V between times t_1 and t_2 is given by $AUC_V := \int_{t_1}^{t_2} V(t)dt = \frac{1}{c}[\frac{p}{\delta}(U(t_1) - U(t_2) + I(t_1) - I(t_2)) + V(t_1) - V(t_2)]$. Therefore, assuming $U(t_1) = U(t_0)$, $I(t_1) = I(t_0)$, $V(t_1) = V(t_0)$, with $U(t_0) \gg I(t_0)$, $U(t_0) \gg V(t_0)$, and $U(t_2) = U_\infty$, $I(t_2) = 0$ and $V(t_2) = 0$, which gives $AUC_V \approx \frac{1}{c}[\frac{p}{\delta}(U(t_0) - U_\infty)]$. This way, if U_∞ is increased with respect to the value corresponding to the untreated case, the area under the curve of viral load decreases. Moreover, as shown in [17], the viral load at time to peak is monotonically decreasing with antiviral therapy reducing β or p.

2.4 Inclusion of PK and PD of antiviral treatment

The idea now is to formally incorporate the PD and PK of antivirals into system (2.1) to obtain a controlled system, i.e., a system with certain

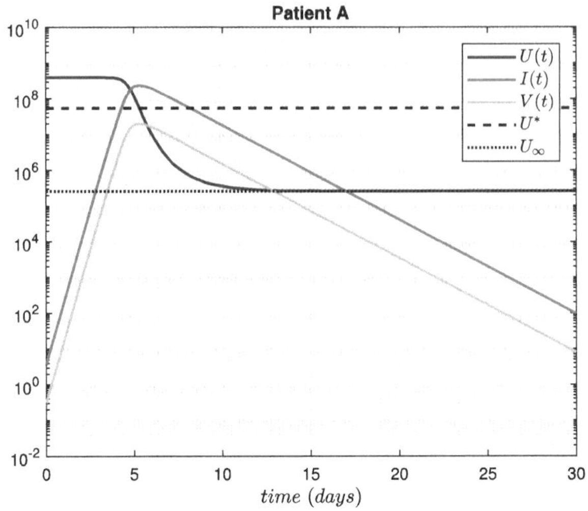

FIGURE 2.4 Time evolution of virtual patient A. As $c \gg \delta$ (as is always the case for real patient data), $I(t) \approx \frac{c}{p} V(t)$ for all $t \geq 0$.

control actions – given by the antivirals – that allows us to (even partially) modify the whole system dynamics according to some control objectives. In contrast to vaccines that kill the virus, antivirals just reduce the viral infection and replication rates, hence reducing the advance of the infection in the respiratory tract. The PD is introduced in system (2.1) as follows:

$$\dot{U}(t) = -\beta(1-\eta(t))U(t)V(t), \tag{2.10a}$$

$$\dot{I}(t) = \beta(1-\eta(t))U(t)V(t) - \delta I(t), \tag{2.10b}$$

$$\dot{V}(t) = pI(t) - cV(t), \tag{2.10c}$$

where $\eta(t) \in [0, 1)$ represents the inhibitory antiviral effects affecting the infection rate β (note that, according to [17], the effect of antivirals on the replication rate p is analogous to the one on β, since both parameters affect the reproduction number $\mathcal{R}$ in the same way).

On the other hand, the PK is modeled as a one compartment with an impulsive input action (to properly account for pill intake or injections):

$$\dot{D}(t) = -\delta_D D(t), \quad t \neq t_k, \tag{2.11a}$$

$$D(t_k) = D(t_k^-) + u_{k-1}, \quad k \in \mathbb{I}, \tag{2.11b}$$

where D is the amount of drug available (with $D(0) = D_0 = 0$), δ_D is the drug elimination rate, and the antiviral dose u_k enters the system impulsively at times $t_k := kT$, with $T > 0$ being a fix time interval and $k \in \mathbb{I}$. Time t_k^- denotes the time just before t_k, i.e., $D(t_k^-) = \lim_{\delta \to 0^+} D(t_k - \delta)$. Note that

(2.11) is an impulsively controlled continuous-time system, which shows discontinuities of the first kind (jumps) at times t_k and free responses in $t \in [t_k, t_{k+1})$ (see [27] for details).

Finally, the way the drug D enters system (2.10) is by means of η as follows:

$$\eta(t) = \frac{D(t)}{D(t) + EC_{50}}, \tag{2.12}$$

where EC_{50} represents the drug concentration in the blood when the drug is half-maximal; $\eta(t)$ is assumed to be in $[0, \eta_{\max})$, with $\eta_{\max} < 1$ (no full antiviral effect is considered, since this is an unrealistic scenario).

2.4.1 Impulsive scheme

Based on the previous PK and PD analysis, the complete COVID-19 infection model taking into account antiviral treatment (the controlled system) reads as follows:

$$\dot{U}(t) = -\beta(1 - \eta(t))U(t)V(t), \quad t \neq t_k, \tag{2.13a}$$
$$\dot{I}(t) = \beta(1 - \eta(t))U(t)V(t) - \delta I(t), \quad t \neq t_k, \tag{2.13b}$$
$$\dot{V}(t) = pI(t) - cV(t), \quad t \neq t_k, \tag{2.13c}$$
$$\dot{D}(t) = -\delta_D D(t), \quad t \neq t_k, \tag{2.13d}$$
$$D(t_k) = D(t_k^-) + u_{k-1}, \quad k \in \mathbb{I}, \tag{2.13e}$$

with initial conditions given by $x_0 = (U_0, I_0, V_0, D_0)$. Given that $D(t) \geq 0$ for all $t \geq 0$, the constraint set $\mathcal{X}$ is enlarged to be $\tilde{\mathcal{X}} := \mathbb{R}^4_{\geq 0}$. Also, a constraint for the input, u, is defined as $\tilde{\mathbb{U}} := \{u \in \mathbb{R} : 0 \leq u \leq u_{\max}\}$, where $u_{\max}$ represent the maximal antiviral dosage ($u_{\max}$ is usually determined by the drug side effects and maximal effectiveness, $0 \leq \eta_{\max} < 1$), while set $\mathcal{X}$ is enlarged by considering $\tilde{\mathcal{X}} := \mathcal{X} \times \mathbb{R}_{\geq 0}$. A detailed study of the stability of impulsive systems can be found in [37].

2.4.2 Simulation example

We resume here the simulation of the virtual patient A to demonstrate the impulsive control actions describing the effects of antiviral administration. It is assumed that antivirals affect the infection rate β, while the initial condition for D is $D_0 = 0$, $\delta_D = 2$ (days^{-1}), and $EC_{50}^p = 75$ (mg). A scenario of 30 days was simulated, and a permanent dose of $u_k = 20$ (mg) of antivirals is administered each T days, starting at $t_i = 4$ days, with $T = 1$, $T = 2$, and $T = 0.5$. As shown in Figs. 2.5a–2.5b, the system response is quite different for different sampling times. For $T = 1$ days, the antiviral treatment is able to decrease V from the beginning. On the other hand, for $T = 2$, the

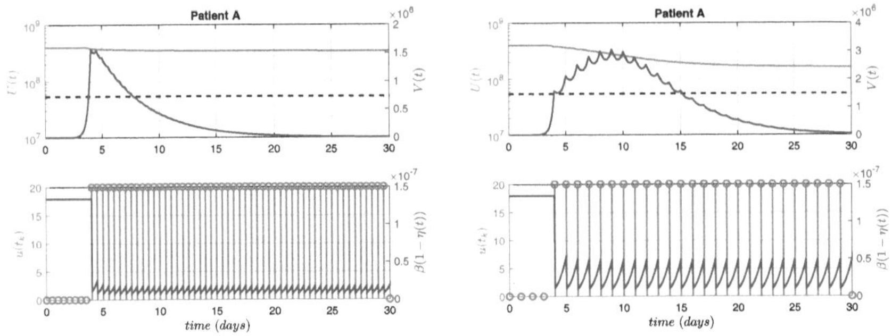

(a) (b)

FIGURE 2.5 (a) Time evolution of virtual patient A, with $u = 20$ mg of antivirals and $T = 1$ days. (b) Time evolution of virtual patient A, with $u = 20$ mg of antivirals and $T = 2$ days.

treatment is unable to stop the spread of the virus (V continues increasing after the treatment is initiated), as shown in Fig. 2.5b. Clearly, the effect of larger values of T is equivalent to smaller values of the dose u_k. In what follows, for the sake of clarity, T will be fixed in 1 day and only the (constant) value of the doses u_k – together with the initial and final time of the treatment – will be modified to analyze the different outcomes.

2.5 Control strategies to tailor therapies

Control objectives in "in-host" infections can be defined in several ways. The peak of the virus load uses to be a critical index to minimize, since it is directly related to the severity of the infection and the ineffective capacity of the host. However, other indexes – usually put in second place – are also important. This is the case of the time the infection lasts in the host over significant levels [17] – including virus rebounds after reaching a pseudosteady state, and the total viral load or infected cells at the end of the infection (i.e., the AUC of V and I). These latter indexes also inform (in a different manner) about the severity of the infection and the time during which the host is able to infect other individuals and are directly determined by the amount of susceptible cells at the end of the infection. So the twofold control objective is defined as follows.

Definition 2.14 (Control objectives). The control objective for the closed-loop (2.13) consists in both maximizing the final value of susceptible/uninfected cells at the end of the infection, U_∞, and minimizing the virus peak, $\hat{V}$. We denote these objectives as Objectives 1 and 2, respectively.

As said in the previous section, antivirals affect the infection rate β, by the time-variant factor $(1 - \eta(t))$. Accordingly, the reproduction number $\mathcal{R}$

will be also time-varying, following the formula

$$\mathcal{R}(t) := \frac{\beta(1-\eta(t))p}{c\delta}, \tag{2.14}$$

and the original reproduction number – i.e., the one corresponding to no treatment – will be denoted as $\mathcal{R}(0)$ for clarity ($\mathcal{R}(0)$ is the reproduction number at the outbreak of the infection, when $u_0 = 0$ and $\eta = 0$).

We will assume in the following a single interval antiviral treatment, consisting in a single fixed dose of antiviral, applied during a finite period of time. At the outbreak of the infection ($t = 0$), it is $(U(0), I(0), V(0)) := (U_0, 0, \epsilon)$, with $\epsilon > 0$ arbitrary small. Then, the single interval treatment is defined by the following input function:

$$u_k = u(t_k) = \begin{cases} 0 & \text{for } t_k \in [0, t_i), \\ u_i & \text{for } t_k \in [t_i, t_f], \\ 0 & \text{for } t_k \in (t_f, \infty), \end{cases} \tag{2.15}$$

where $t_i < \hat{t}(\mathcal{R}(0))$, with $\hat{t}(\mathcal{R}(0))$ being the time of the peak of $V(\tau)$ when no treatment is implemented, $u_i \in [0, u_{\max}]$, and $t_f > t_i$, but finite. Note that after t_f, $u(t_k) = 0$, which means that $\eta(t) \to 0$ and $\mathcal{R} \to \mathcal{R}(0)$.

2.5.1 First control objective: maximizing the final value of the uninfected cells

The control problem we want to solve first reads as follows. For a given initial time, $t_i < \hat{t}$, find u_i (which has an associated $\mathcal{R}_i$ and η_i) and t_f (finite) to maximize $U_\infty = U_\infty(\mathcal{R}(0), U(t_f), I(t_f), V(t_f))$. This control problem accounts for the first control objective; next, some comments will be made concerning the second one.

A critical point concerning antiviral treatments – that is usually disregarded – is that they are always transitory control actions, not permanent ones. It is not possible to maintain a given treatment for a long time, and its interruption must be explicitly considered in any antiviral schedule.

So, according to the stability results from the previous sections, the following property holds.

Property 2.15 (Upper bound for U_∞). *Consider system (2.13) with $U(0) > U^*$ (or $U(0)\mathcal{R}(0) > 1$). No matter which kind of antiviral treatment is implemented at time t_i, if it is interrupted at some finite time $t_f > t_i$ (as is always the case), the system converges to an equilibrium state $(U_\infty, 0, 0)$ with $U_\infty \leq U^*$, with U^* being the critical value for U corresponding to no antiviral treatment, i.e., $U^* = 1/\mathcal{R}(0)$.*

Proof. We proceed by contradiction. Assume that $U_\infty > U^*$. Consider system (2.10) for $t \geq t_f$. Since the antiviral treatment is interrupted at time

t_f, $\eta(t) \searrow 0$, for $t \geq t_f$. By Eq. (2.14) we have $\mathcal{R}(t) \nearrow \mathcal{R}(0)$ for $t \geq t_f$. If we denote $U_c(t) = 1/\mathcal{R}(t)$ we have $U_c \searrow U^*$ for $t \geq t_f$. Since $U_\infty > U^*$, $U_\infty > U_c(t)$ for $t > T$, with T large enough. Hence $(U(t), I(t), V(t))$ is converging to an equilibrium point in the unstable part, which is a contradiction. Therefore $U_\infty \leq U^*$, which concludes the proof. □

Remark 2.16. From a clinical perspective, what Property 2.15 establishes is more than a simple upper bound for U_∞. It says that the best an antiviral treatment can do in terms of the *total amount of virus (or infected cells) at the end of the infection*, $V_{\text{tot}} := \int_{t=0}^{\infty} V(t)dt \approx \int_{t=0}^{\infty} \frac{p}{c} I(t)dt$, is to reach a minimal value intrinsically determined by the system parameters ($\mathcal{R}(0)$). Furthermore, the instantaneous *peak of* $V(t)$, for $t > 0$, which is the other critical index for the severity of the infection (whose minimization is the second control objective), is independent of the latter lower limit (as shown later on) and can be minimized while maintaining V_{tot} at its minimal value. This represents a new paradigm concerning what (and what not) antiviral treatments can do in acute infections.

In the search of such a value, the next definition is stated.

Definition 2.17 (Goldilocks antiviral dose). The Goldilocks antiviral dose (GAD), $u^g = u^g(t_i)$, is the one that, if applied at $t_i < \hat{t}(\mathcal{R}(0))$, produces $U_\infty(\mathcal{R}^g, U(t_i), I(t_i), V(t_i)) = U^*$, where $\mathcal{R}^g$ is determined by u^g, at steady state.

Remark 2.18 (u^g computation). Given t_i and $\mathcal{R}(0)$, $u^g = u^g(U(t_i), I(t_i), V(t_i)) = u^g(t_i)$ can be obtained numerically by means of Algorithm 2.1.

Algorithm 2.1: Computation of $u^g(t_i)$

$u_k = 0$, $U^* = 1/\mathcal{R}(0)$;
Compute U_i, I_i, and V_i by integrating system (2.1) from 0 to t_i, starting at (U_0, I_0, V_0);
Compute $U(t_f)$ by integrating system (2.13) from t_i to t_f, starting at (U_i, I_i, V_i), with u_k;
while $U(t_f) <= U^*$ **do**
  $u_k = u_k + 0.001$;
  Compute $U(t_f)$ by integrating system (2.13) from t_i to t_f, starting at (U_i, I_i, V_i), with u_k;
end
$u^g = u_k$;

Clearly, Goldilocks antiviral treatment cannot be applied indefinitely, since t_f is finite. However, it can be applied up to a time t_f large enough such that $(U(t_f), I(t_f), V(t_f))$ is arbitrarily close to $(U^*, 0, 0)$ from above.

This latter scenario is denoted as quasisteady state (QSS), and it allows us to introduce the following definition.

Definition 2.19 (Quasioptimal single interval antiviral treatment). Consider a given starting time, $t_i \in (0, \hat{t}(\mathcal{R}(0)))$. Then the quasioptimal single interval antiviral treatment consists in applying u^g, up to a time t_f large enough for the system to reach a QSS condition (i.e., $U(t_f) \approx U^*$ $I(t_f) \approx 0$, $V(t_f) \approx 0$).

Remark 2.20. Clearly, the latter definition refers to a quasioptimal single interval control action, because larger values of t_f will produce values of $U(t_f)$, $I(t_f)$, and $V(t_f)$ closer to U^*, 0, and 0, respectively, so $U_\infty(\mathcal{R}(0), U(t_f), I(t_f), V(t_f))$ will be closer to U^*.

The next theorem, which is one of the main contributions of the work, summarizes the latter results by means of a classification that considers every possible single interval treatment case.

Theorem 2.21 (Single interval antiviral treatment scenarios). *Consider system (2.1) with initial conditions $(U(0), I(0), V(0)) = (U_0, 0, \epsilon)$, with $\epsilon > 0$ arbitrary small, and $\mathcal{R}(0)$ such that $U(0) > U^*$. Consider also single interval antiviral treatment (as the one defined in (2.15)), with a given starting time $t_i \in (0, \hat{t}(\mathcal{R}(0)))$ and a finite final time t_f. Define soft and strong treatments depending on if $u_i < u^g$ or $u_i > u^g$, respectively. Define also long- and short-term treatments depending on if the system reaches or does not reach a QSS at t_f. Then, the following scenarios can take place:*

i. *Quasioptimal single interval antiviral treatment: if $u_i = u^g$ and t_f is such that $(U(t_f), I(t_f), V(t_f))$ reaches a QSS, then $U_\infty(\mathcal{R}(0), U(t_f), I(t_f), V(t_f)) \approx U^*$. Furthermore, the closer $U(t_f)$ is to U^* (or $I(t_f)$ and $V(t_f)$ to zero), the closer $U_\infty(\mathcal{R}(0), U(t_f), I(t_f), V(t_f))$ will be to U^*.*

ii. *Soft long-term antiviral treatment: if $(U(t_f), I(t_f), V(t_f))$ reaches a QSS and $U(t_f) < U^*$, then $U_\infty(\mathcal{R}(0), U(t_f), I(t_f), V(t_f)) \approx U(t_f) < U^*$; i.e., $U(t)$ will remain approximately constant for $t \geq t_f$. Furthermore, the softer a soft long-term antiviral treatment is, the smaller will be $U_\infty(\mathcal{R}(0), U(t_f), I(t_f), V(t_f))$.*

iii. *Strong long-term antiviral treatment: if $(U(t_f), I(t_f), V(t_f))$ reaches a QSS and $U(t_f) > U^*$, a* rebound *will necessarily take place at some time $\hat{\hat{t}} > t_f$ and, finally, the system will converge to a $U_\infty(\mathcal{R}(0), U(t_f), I(t_f), V(t_f)) < U^*$. Furthermore, the stronger a strong long-term antiviral treatment is, the larger will be the rebound and the smaller will be $U_\infty(\mathcal{R}(0), U(t_f), I(t_f), V(t_f))$.*

iv. *Short-term antiviral treatment: if $(U(t_f), I(t_f), V(t_f))$ does not reach a QSS (i.e., if $V(t_f) \not\approx 0$), then soft, strong, and Goldilocks doses will necessarily produce values of $U_\infty(\mathcal{R}(0), U(t_f), I(t_f), V(t_f))$ significantly smaller than the one obtained by quasioptimal single interval treatment. In general, larger values of $V(t_f)$ will produce smaller values of $U_\infty(\mathcal{R}(0), U(t_f),$*

$I(t_f), V(t_f))$. This case includes the particular case where the treatment is interrupted at the very moment at which $U(t_f) = U^*$, but with $V(t_f) \not\approx 0$. This means that the critical value of U needs to be reached as a steady state, not as a transitory one.

Proof. The proof follows from the stability results shown in Sections 2.3 and 2.3.6:

i. Given that $u_i = u^g$ is implemented for $t \in [t_i, t_f]$, t_f is finite but large enough and $U_\infty(\mathcal{R}^g, U(t_i), I(t_i), V(t_i)) = U^*$, then $U(t_f)$ approaches U^* and $V(t_f)$ approaches zero, from above, as t_f increases. This means that at t_f, when the treatment is interrupted, $(U(t_f), I(t_f), V(t_f))$ is close to the unstable equilibrium set $\mathcal{X}_s^{un}$. Then, by Property 2.12(ii), function $U_\infty(\mathcal{R}(0), U(t_f), I(t_f), V(t_f))$ is such that the closer $(U(t_f), I(t_f), V(t_f))$ is to the equilibrium point $(U^*, 0, 0)$, with $U(t_f) > U^*$, the closer will $U_\infty(\mathcal{R}(0), U(t_f), I(t_f), V(t_f))$ be to U^*, with $U_\infty < U^*$ (see the "pine" shape of U_∞ around U^*, for $I \approx 0$, in Fig. 2.3). Indeed, by the ϵ-δ stability of the equilibrium state $(U^*, 0, 0)$, for each (arbitrary small) $\epsilon > 0$, if there exists $\delta > 0$ such that if the system starts in a ball of radius δ centered at $(U^*, 0, 0)$, it will keep indeterminately in the ball of radius ϵ centered at $(U^*, 0, 0)$. Furthermore, it is possible to define invariant sets around $(U^*, 0, 0)$ by considering the level sets of the Lyapunov function (2.9), with $\bar{U} = U^*$, or even the level sets of function $J(U, I, V) := U^* - U_\infty(\mathcal{R}, U, I, V)$, with a fixed $\mathcal{R} > 0$. This way, once the system enters any arbitrary small level set of the latter functions, it cannot leave the set anymore (see Fig. 2.6).

ii. Given that $(U(t_f), I(t_f), V(t_f))$ approaches a steady state with $U(t_f) < U^*$, $(U(t_f), I(t_f), V(t_f))$ is close to the stable equilibrium set $\mathcal{X}_s^{st}$ when the treatment is interrupted. Then the system will converge to an equilibrium with $U_\infty(\mathcal{R}(0), U(t_f), I(t_f), V(t_f))$ close to $U(t_f)$. Softer antiviral treatment produces smaller values of $U(t_f)$, and by Property 2.12(iii), smaller values of $U(t_f)$ produce smaller values of $U_\infty(\mathcal{R}(0), U(t_f), I(t_f), V(t_f))$.

iii. Given that $(U(t_f), I(t_f), V(t_f))$ approaches a steady state with $U(t_f) > U^*$, $(U(t_f), I(t_f), V(t_f))$ is close to the unstable equilibrium set, $\mathcal{X}_s^{un}$, when the treatment is interrupted. Then the system will converge to an equilibrium in the stable equilibrium set, $\mathcal{X}_s^{st}$, with $U_\infty(\mathcal{R}(0), U(t_f), I(t_f), V(t_f)) < U^*$. Stronger antiviral treatment produces greater values of $U(t_f)$, and by Property 2.12(ii), values of $U(t_f)$ further from U^*, from above, produce values of $U_\infty(\mathcal{R}(0), U(t_f), I(t_f), V(t_f))$ further from U^*, from below, when $U(t_f)$ is significantly greater than U^*, no matter how large t_f is and how small $V(t_f)$ is. Note that as long as t_f is finite, $(U(t_f), I(t_f), V(t_f))$ cannot reach $\mathcal{X}_s^{un}$, so $V(t_f)$, even when arbitrarily small, is greater than zero.

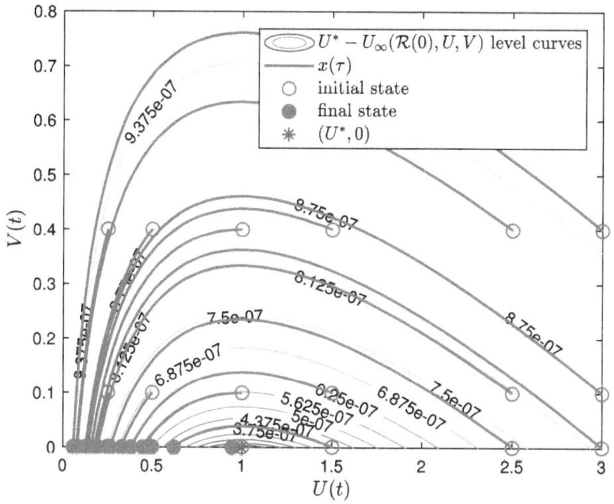

FIGURE 2.6 Phase portrait for system (2.1) ($\beta = 1/2$, $\delta = 1/5$, $p = 2$, $c = 5$) in the U, V plane ($I = (c/p)V$), for different starting points (red (gray in print version) lines) and level curves of function $J(U, I, V) := U^* - U_\infty(\mathcal{R}, U, I, V)$, $I = (c/p)V$. Function J is positive for all $(U, V) \neq (U^*, 0)$ and is null at $(U^*, 0)$, and $\dot{J}(U, V) = 0$ along the solution of system (2.1) (since $U_\infty(\mathcal{R}, U, I, V)$ is). So its level sets are arbitrarily small invariant sets around $(U^*, 0)$. Note that starting states close to $(U^*, 0)$ produce time evolution close to U^*, as they determine the ϵ-δ stability.

So, once the treatment is interrupted, the system evolves to an equilibrium in $\mathcal{X}_s^{un}$. The system will evolve to an equilibrium in $\mathcal{X}_s^{st}$, with $U_\infty(\mathcal{R}(0), U(t_f), I(t_f), V(t_f))$ significantly smaller than U^*. Furthermore, to go from $U(t_f)$ to $U_\infty(\mathcal{R}(0), U(t_f), I(t_f), V(t_f))$, for $t > t_f$, the system significantly increases $V(t)$, and this effect is known as a rebound.

iv. Given that $(U(t_f), I(t_f), V(t_f))$ is a transitory state, it does not approach any equilibrium. This means that $V(t_f)$ is significantly greater than 0, and according to Lemma 2.7, the maximum of $U_\infty(\mathcal{R}(0), U(t_f), I(t_f), V(t_f))$ over $\Omega(\varepsilon) = \{(U, I, V) \in \mathbb{R}^3_{\geq 0} : I \geq \varepsilon, V \geq \varepsilon\}$ is given by $-W(-\mathcal{R}U^* e^{-\mathcal{R}(U^* + \varepsilon + \frac{\delta}{p}\varepsilon)})/\mathcal{R}$, which is a decreasing function of ε and reaches U^* only when $\varepsilon = 0$ (see Fig. 2.3). Then, independently of the value of $U(t_f)$, $U_\infty(\mathcal{R}(0), U(t_f), I(t_f), V(t_f))$ will be (maybe significantly) smaller than the one obtained with quasioptimal single interval treatment, in which $\varepsilon \approx 0$. □

2.5.2 Second control objective: minimizing the virus peak

The quasioptimal single interval treatment clearly accounts for a steady-state condition, given that any realistic treatment needs to be interrupted at a finite time. Furthermore, given that only one antiviral dose, u_i, is considered for the treatment, once the quasioptimal single interval treatment is determined ($u_i = u^g$), so is the peak (maximum over t) of the virus, $\hat{V}$, i.e., there is a unique $\hat{V}$ for each single interval control action, u_i.

However, if a more general control action is considered, in such a way that u_k assume several values in the interval from t_i to t_f, $\hat{V}$ can be arbitrarily reduced. Indeed, given that U_∞ depends only on the fact that $U \approx U^*$ and $V \approx 0$ at t_f, then stronger antiviral doses can be used at the beginning of the treatment to lower the peak of V. If for instance two consecutive single interval control actions are implemented – the first one with a high dose, applied from t_i to t_1, and the second one with the quasioptimal antiviral, $u^g(t_1)$, applied from t_1 to a large enough t_f – a lower peak of V will necessarily be obtained in contrast to one corresponding to the quasioptimal single interval control.

Although this chapter is not devoted to analyze control strategies different from the single interval one, it is worth to remark this latter point since it states that: (1) both control objectives are independent, in the sense that if a given upper bound for V is stated from the beginning (to avoid complication and/or to reduce the infectivity of the host) it is in general possible to design control strategies that both make $V(t)$ not overpass the upper bound and make $U_\infty \approx U^*$, and (2) the entire concept of a maximum or peak for V, for a given treatment, has sense only when $U_\infty \leq U^*$, since otherwise, a rebound of the virus will occur once the treatment is interrupted, and a new peak for V may be reached.

Figs. 2.11a and 2.11b, in the Simulation section, show an example of a two-step interval treatment that produces a peak of V smaller than the one corresponding to the quasioptimal single interval one.

2.6 Simulation results

In this section each of the cases of Theorem 2.21 and the case of two-step interval control action of Section 2.5.2 are simulated for data coming from patient A, introduced in Sections 2.3.7 and 2.4.2. As already said, $\delta_D = 2$ (days^{-1}), $EC_{50}^p = 75$ (mg), and the sampling time is selected to be $T = 1$ day. Initial conditions are given by $(U_0, I_0, V_0) = (4 \times 10^8, 0, 0.31)$. Also, recall that $U^* = 5.44 \times 10^7$ and the untreated peak of V is given by $\hat{V} = 1.98 \times 10^7$.

2.6.1 Strong long-term treatment. Virus rebound

Fig. 2.7a shows the time evolution of U (logarithmic scale), V, and u_k for patient A, when strong long-term antiviral treatment is implemented. The treatment starts at $t_i = 4$ days and finishes at $t_f = 30$ days, while several strong doses are administered: $u_i = [21, 25, 35]$ mg.

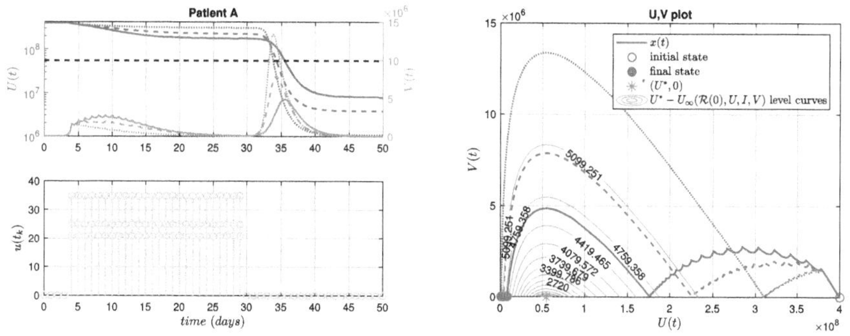

(a) (b)

FIGURE 2.7 (a) Time evolution of virtual patient A, with different doses of antiviral: $u_i = 21$ mg, solid line; $u_i = 25$ mg, dashed line; $u_i = 35$ mg, dotted line. (b) Phase portrait in the U, V space and level curves of the Lyapunov function $J(U, I, V) := U^* - U_\infty(\mathcal{R}(0), U, I, V)$, around $(U^*, 0)$.

As can be seen, at t_f the value of U is greater than U^* while $V \approx 0$, so the viral load V rebounds after some time, producing a second (and larger) peak. More important, U_∞ ends up at a value significantly smaller than U^*. The values of U_∞ and $\hat{V}$ corresponding to the three doses are given by $U_\infty = [7.98 \times 10^6, 3.79 \times 10^6, 1.07 \times 10^5]$ and $\hat{V} = [4.84 \times 10^6, 7.86 \times 10^6, 1.34 \times 10^7]$, respectively.

To have a better idea of how the system behaves around state $(U^*, 0)$, Fig. 2.7b shows the phase portrait in the space U, V, together with the level curves of the Lyapunov function $J(U, I, V) := U^* - U_\infty(\mathcal{R}(0), U, I, V)$. At time t_f, when the treatment is interrupted, $V(t_f) \approx 0$ and $U(t_f) > U^*$, so the system is close to an unstable equilibrium point. So, for $t > t_f$ the state is attracted to an equilibrium in the AS equilibrium set $\mathcal{X}_s^{st}$, following outer level curves of J. Outer level curves of J means both a small U_∞ and a large $\hat{V}$.

2.6.2 Soft long-term treatment

Fig. 2.8a shows the time evolution of U (logarithmic scale), V, and u_k, when soft long-term antiviral treatment is implemented. The treatment starts at $t_i = 4$ days and finishes at $t_f = 30$ days, while several soft doses are administered: $u_i = [4, 6, 8]$ mg.

As can be seen, at t_f the value of U is smaller than U^*, while $V \approx 0$, so the viral load V decreases after the treatment is interrupted. The values of U_∞ and $\hat{V}$ corresponding to the three doses are given by $U_\infty = [8.98 \times 10^6, 1.90 \times 10^7, 3.25 \times 10^7]$ and $\hat{V} = [1.36 \times 10^7, 1.16 \times 10^7, 9.70e \times 10^6]$.

Fig. 2.8b shows the phase portrait in the space U, V, together with the level curves of the Lyapunov function $J(U, I, V)$. At time t_f, when the

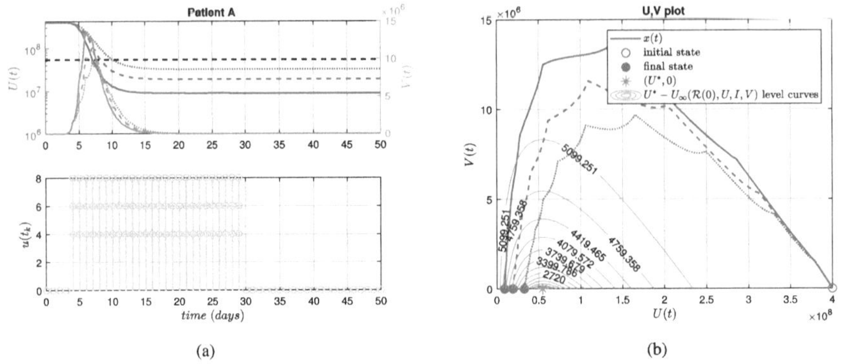

FIGURE 2.8 (a) Time evolution of virtual patient A, with different doses of antiviral: $u_i = 4$ mg, solid line; $u_i = 6$ mg, dashed line; $u_i = 8$ mg, dotted line. (b) Phase portrait in the U, V space and level curves of the Lyapunov function $J(U, I, V) := U^* - U_\infty(\mathcal{R}(0), U, I, V)$, around $(U^*, 0)$.

treatment is interrupted, $V(t_f) \approx 0$ and $U(t_f) < U^*$, so the system is close to a stable equilibrium point. So, for $t > t_f$ the state remains almost unmodified.

2.6.3 Quasioptimal single interval treatment

Fig. 2.9a shows the time evolution of U (logarithmic scale), V, and u_k, when the quasioptimal single interval antiviral treatment is administered. The treatment starts at $t_i = 4$ days and finishes at $t_f = 30$ days, while the Goldilocks dose is given by $u_i = u^g(t_i) = 10.5$ mg. The values of U_∞ and $\hat{V}$ are given by $U_\infty = 5.34 \times 10^7$ and $\hat{V} = 7.73 \times 10^6$, respectively.

Fig. 2.9b shows the phase portrait in the space U, V, together with the level curves of the Lyapunov function $J(U, I, V)$. As can be seen, the system follows the only one trajectory that goes directly from (U_0, V_0) to $(U^*, 0)$: any other path goes necessarily to an equilibrium with $U_\infty < U^*$.

2.6.4 Short-term treatment

Fig. 2.10a shows the time evolution of U (logarithmic scale), V, and u_k, when a short-term treatment is implemented. The treatment starts at $t_i = 4$ days and finishes at $t_f = 15$ days, while several doses – smaller and greater than $u^g(t_i)$ – are administered: $u_i = [10, 15, 20, 25]$ mg. The values of U_∞ and $\hat{V}$ corresponding to the four doses are given by $U_\infty = [2.98 \times 10^6, 1.39 \times 10^7, 5.06 \times 10^7, 2.26 \times 10^7]$ and $\hat{V} = [8.05 \times 10^6, 4.98 \times 10^6, 6.67 \times 10^6, 1.01 \times 10^1]$, respectively.

Fig. 2.10b shows the phase portrait in the space U, V. Given that trajectories go along the level curves of the Lyapunov function $J(U, I, V)$, any

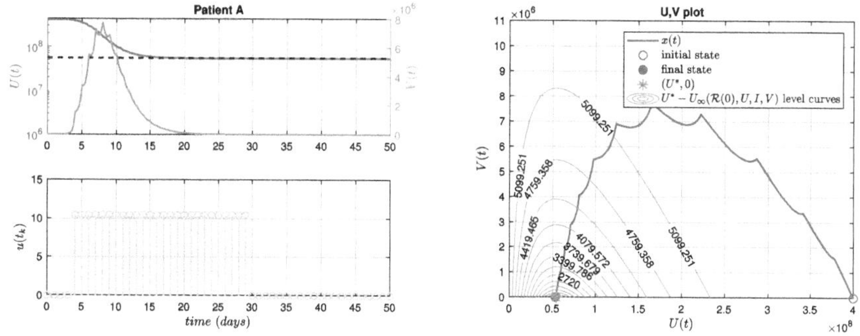

(a) (b)

FIGURE 2.9 (a) Time evolution of virtual patient A, with $u_i = u^g = 10.5$ mg of antiviral. (b) Phase portrait in the U, V space and level curves of the Lyapunov function $J(U, I, V) := U^* - U_\infty(\mathcal{R}(0), U, I, V)$, around $(U^*, 0)$.

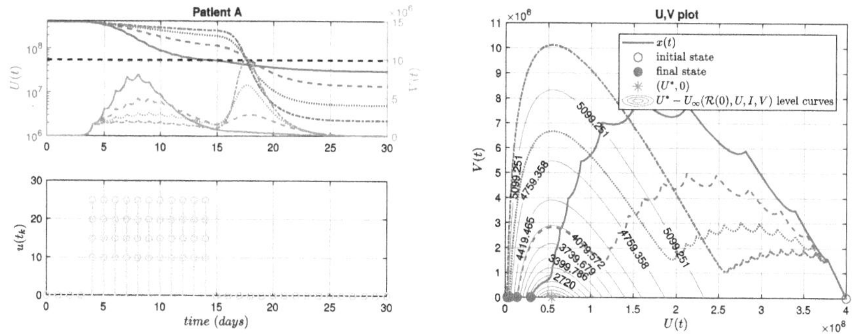

(a) (b)

FIGURE 2.10 (a) Time evolution of virtual patient A, with different doses of antiviral: $u_i = 10$ mg, solid line; $u_i = 15$ mg, dashed line; $u_i = 20$ mg, dotted line; and $u_i = 25$ mg, dashed-dotted line. (b) Phase portrait in the U, V space and level curves of the Lyapunov function $J(U, I, V) := U^* - U_\infty(\mathcal{R}(0), U, I, V)$, around $(U^*, 0)$.

short term treatment – i.e., producing $V(t_f) \not\approx 0$ – will make the system surround the state $(U^*, 0)$ by an outer level curve, thus finishing at some U_∞ significantly smaller than U^*. As before, outer level curves of J means both a small U_∞ and a large $\hat{V}$.

2.6.5 Two-step treatment, lowering the peak of V

Finally, a scenario is simulated to show that it is always possible to lower the peak of V – while maintaining $U_\infty \approx U^*$ – if a control sequence more complex than the single interval one is implemented. Fig. 2.11a

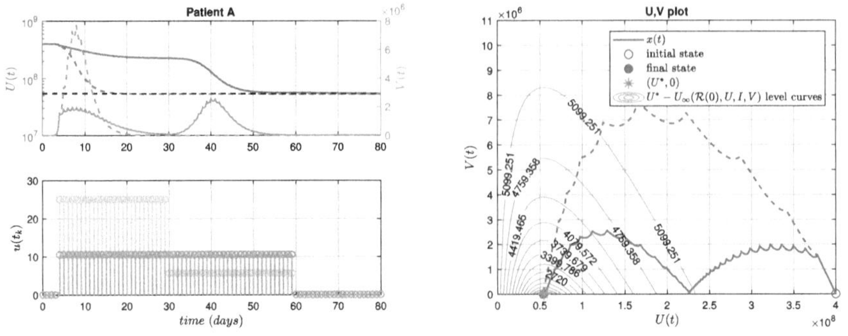

(a) (b)

FIGURE 2.11 (a) Time evolution of virtual patient A, with different doses of antiviral: $u_i = 25$ mg, from $t_i = 4$ to $t_m = 30$ days; and $u_i = u^g(t_m) = 5.6$ mg, from $t_m = 30$ to $t_f = 60$ days (input, yellow (light gray in print version) stems; U and V in solid lines). In the dashed line the progression of target cells and viral particles is plotted considering the quasioptimal single interval treatment, given by $u_i = u^g(t_i) = 10$ mg from $t_i = 4$ to $t_f = 60$ (green (mid gray in print version) stems). (b) Phase portrait in the U, V space.

shows the time evolution of U (solid blue (dark gray in print version) line, logarithmic scale) and V (solid red (gray in print version) line) corresponding to a two-step interval control, the first step consisting in $u_i = 25$ mg from $t_i = 4$ to $t_m = 30$ days and the second one consisting in $u_i = u^g(t_m) = 5.6$ mg from $t_m = 30$ to $t_f = 60$ (solid line). Also, the quasioptimal single interval control of Section 2.6.3 is shown to compare the performance (dashed line). As can be seen, the peak of V is significantly reduced, from $\hat{V} = 7.73 \times 10^6$ to $\hat{V} = 2.57 \times 10^6$, while U_∞ is almost the same in both cases. Fig. 2.11b shows the phase portraits of the two control strategies (solid line, two-step control; dashed line, single interval control), where also the reduction of the virus peak can be seen. This simple two-step strategy shows that with a more sophisticated control strategy (i.e., by means of a proper optimal control formulation) the virus peak can be arbitrarily reduced, maintaining the condition $U_\infty \approx U^*$. This is indeed a matter of future research.

2.7 Conclusions and future works

In this work, the stability and general long-term behavior of UIV-type models have been fully analyzed. A quasioptimal control action – consisting in the finite-time single interval antiviral treatment producing the minimal possible final amount of dead cells – was found. The analysis shows also that more complex control strategies can account for both control objectives simultaneously: minimizing the virus peak, while keeping the final amount of dead cells at its maximum. A detailed analysis of suboptimal scenarios permits to enumerate the following main results:

i. To apply soft antiviral treatment during a long time (even no treatment at all), expecting the noninfected cells would evolve alone to the critical value U^*, is not an option. Open-loop U_∞ is in general significantly smaller than U^* (particularly for the reported values of $\mathcal{R}$ for COVID-19).

ii. To apply strong antiviral treatment for a long time expecting the virus will die out alone is not an option. Strong antivirals produce values of U at the end of treatment larger than U^*, but these final values are artificially stable steady states, since once the treatment is interrupted or reduced, a virus rebound will necessarily occur at some future time and U_∞ will be significantly smaller than U^*.

iii. To apply any antiviral treatment (soft or strong) for a short period of time such that the system is not able to reach a QSS (i.e., when V at the end of the treatment is not close to zero) is not an option. If the treatment is interrupted at a transient state, the initial conditions for the next time period are such that U_∞ will be significantly smaller than U^*.

iv. According to the latter results, the best option is to apply an antiviral treatment such that the system reaches a QSS with $U \approx U^*$ and $V \approx 0$ at the end of the treatment. This is what we call the "quasioptimal single interval antiviral treatment," since it makes the system approach the maximal final value of uninfected cells ($U_\infty \approx U^*$), without infection rebounds.

v. An important point to be made is that the quasioptimal single interval antiviral treatment does not determine the peak of the virus. Quasioptimal conditions for U_∞ are stationary, while conditions for minimizing $\hat{V}$ are transitory, so both objectives can be accounted for simultaneously.

Future works include the study of more complex control strategies (mainly model-based control strategies such as MPC) and the explicit consideration of time-varying immune systems.

References

[1] A.S. Perelson, D.E. Kirschner, R. De Boer, Dynamics of HIV infection of CD4+ T cells, Mathematical Biosciences 114 (1) (1993) 81–125.

[2] M. Legrand, E. Comets, G. Aymard, R. Tubiana, C. Katlama, B. Diquet, An in vivo pharmacokinetic/pharmacodynamic model for antiretroviral combination, HIV Clinical Trials 4 (3) (2003) 170–183.

[3] A.S. Perelson, R.M. Ribeiro, Modeling the within-host dynamics of HIV infection, BMC Biology 11 (1) (2013) 96.

[4] S.M. Ciupe, R.M. Ribeiro, P.W. Nelson, A.S. Perelson, Modeling the mechanisms of acute hepatitis b virus infection, Journal of Theoretical Biology 247 (1) (2007) 23–35.

[5] E. Herrmann, A.U. Neumann, J.M. Schmidt, S. Zeuzem, Hepatitis C virus kinetics, Antiviral Therapy 5 (2) (2000) 85–90.
[6] A.U. Neumann, N.P. Lam, H. Dahari, D.R. Gretch, T.E. Wiley, T.J. Layden, A.S. Perelson, Hepatitis C viral dynamics in vivo and the antiviral efficacy of interferon-α therapy, Science 282 (5386) (1998) 103–107.
[7] L. Canini, A.S. Perelson, Viral kinetic modeling: state of the art, Journal of Pharmacokinetics and Pharmacodynamics 41 (5) (2014) 431–443.
[8] P. Baccam, C. Beauchemin, C.A. Macken, F.G. Hayden, A.S. Perelson, Kinetics of influenza A virus infection in humans, Journal of Virology 80 (15) (2006) 7590–7599.
[9] A.M. Smith, A.S. Perelson, Influenza A virus infection kinetics: quantitative data and models, Wiley Interdisciplinary Reviews: Systems Biology and Medicine 3 (4) (2011) 429–445.
[10] G. Hernandez-Mejia, A.Y. Alanis, M. Hernandez-Gonzalez, R. Findeisen, E.A. Hernandez-Vargas, Passivity-based inverse optimal impulsive control for Influenza treatment in the host, IEEE Transactions on Control Systems Technology (2019).
[11] R. Nikin-Beers, S.M. Ciupe, The role of antibody in enhancing dengue virus infection, Mathematical Biosciences 263 (2015) 83–92.
[12] R. Nikin-Beers, S.M. Ciupe, Modelling original antigenic sin in dengue viral infection, Mathematical Medicine and Biology 35 (2) (2018) 257–272.
[13] V. Nguyen, S. Binder, A. Boianelli, M. Meyer-Hermann, E.A. Hernandez-Vargas, Ebola virus infection modelling and identifiability problems, Frontiers in Microbiology 6 (05 2015), https://doi.org/10.3389/fmicb.2015.00257.
[14] P. van den Driessche, Reproduction numbers of infectious disease models, Infectious Disease Modelling 2 (3) (2017) 288–303.
[15] A. Murase, T. Sasaki, T. Kajiwara, Stability analysis of pathogen-immune interaction dynamics, Journal of Mathematical Biology 51 (3) (2005) 247–267.
[16] H.L. Smith, P. De Leenheer, Virus dynamics: a global analysis, SIAM Journal on Applied Mathematics 63 (4) (2003) 1313–1327.
[17] P. Abuin, A. Anderson, A. Ferramosca, E.A. Hernandez-Vargas, A.H. Gonzalez, Dynamical characterization of antiviral effects in Covid-19, arXiv preprint, arXiv:2012.15585, 2020.
[18] S.M. Ciupe, J.M. Heffernan, In-host modeling, Infectious Disease Modelling 2 (2) (2017) 188–202.
[19] P. Cao, J.M. McCaw, The mechanisms for within-host influenza virus control affect model-based assessment and prediction of antiviral treatment, Viruses 9 (8) (2017) 197.
[20] J.B. Rawlings, D.Q. Mayne, M. Diehl, Model Predictive Control: Theory, Computation, and Design, Vol. 2, Nob Hill Publishing, Madison, WI, 2017.
[21] F. Blanchini, S. Miani, Set-Theoretic Methods in Control, Springer, 2008.
[22] P. Abuin, A. Anderson, A. Ferramosca, E.A. Hernandez-Vargas, A.H. Gonzalez, Characterization of SARS-CoV-2 dynamics in the host, Annual Reviews in Control (2020).
[23] R. Eftimie, J.J. Gillard, D.A. Cantrell, Mathematical models for immunology: current state of the art and future research directions, Bulletin of Mathematical Biology 78 (10) (2016) 2091–2134.
[24] E.A. Hernandez-Vargas, Modeling and Control of Infectious Diseases in the Host: With MATLAB and R, Academic Press, 2019.
[25] H.M. Dobrovolny, M.B. Reddy, M.A. Kamal, C.R. Rayner, C.A. Beauchemin, Assessing mathematical models of influenza infections using features of the immune response, PLoS ONE 8 (2) (2013) e57088.
[26] J.-M. Vergnaud, I.-D. Rosca, Assessing Bioavailability of Drug Delivery Systems: Mathematical Modeling, CRC Press, 2005.
[27] P.S. Rivadeneira, A. Ferramosca, A.H. González, Control strategies for nonzero set-point regulation of linear impulsive systems, IEEE Transactions on Automatic Control 63 (9) (2018) 2994–3001.

[28] G. Hernandez-Mejia, A. Alanis, E.A. Hernandez-Vargas, Inverse optimal impulsive control based treatment of influenza infection, IFAC-PapersOnLine 50 (2017) 12185–12190, https://doi.org/10.1016/j.ifacol.2017.08.2272.
[29] P.S. Rivadeneira, C.H. Moog, Impulsive control of single-input nonlinear systems with application to HIV dynamics, Applied Mathematics and Computation 218 (17) (2012) 8462–8474.
[30] H. Dahari, A. Lo, R.M. Ribeiro, A.S. Perelson, Modeling hepatitis C virus dynamics: liver regeneration and critical drug efficacy, Journal of Theoretical Biology 247 (2) (2007) 371–381.
[31] H.M. Dobrovolny, R. Gieschke, B.E. Davies, N.L. Jumbe, C.A. Beauchemin, Neuraminidase inhibitors for treatment of human and avian strain influenza: a comparative modeling study, Journal of Theoretical Biology 269 (1) (2011) 234–244.
[32] R. Wölfel, V.M. Corman, W. Guggemos, M. Seilmaier, S. Zange, M.A. Müller, D. Niemeyer, T.C. Jones, P. Vollmar, C. Rothe, et al., Virological assessment of hospitalized patients with COVID-2019, Nature 581 (7809) (2020) 465–469.
[33] E.A. Hernandez-Vargas, J.X. Velasco-Hernandez, In-host Modelling of COVID-19 Kinetics in Humans, medRxiv, 2020.
[34] L. Perko, Differential Equations and Dynamical Systems, Vol. 7, Springer Science & Business Media, 2013.
[35] J. Kelley, General Topology, Dover Books on Mathematics, Dover Publications, 2017.
[36] H.K. Khalil, J.W. Grizzle, Nonlinear Systems, Vol. 3, Prentice Hall, Upper Saddle River, NJ, 2002.
[37] A. D'Jorge, A.L. Anderson, A. Ferramosca, A.H. González, M. Actis, On stability of nonzero set-point for non linear impulsive control systems, arXiv:2011.12085, 2020.

CHAPTER

3

Input-output approaches for personalized drug dosing of antibiotics

Pauline Thémans[a], *Flora Tshinanu Musuamba*[b], *and Joseph J. Winkin*[a]

[a]Namur Institute for Complex Systems (naXys) and Department of Mathematics, University of Namur, Namur, Belgium, [b]Federal Agency for Medicines and Health Products, Brussels, Belgium

3.1 Introduction

Pharmacokinetics (PK) is a particular field of clinical pharmacology that studies the link between the dose of a drug administered to patients and the drug exposure (concentrations) over time. Thanks to mathematical modeling, clinical pharmacology is an interesting and promising field of application of systems and control theory (see, e.g., [1]). Examples reported in the literature include automated anesthesia [1] and the artificial pancreas [9]. They involve closed-loop control strategies that are based on a control system which continuously adjusts the rate of infusion of the drug.

This chapter focuses on antibiotic (AB) treatments given by constant intravenous (IV) infusion at regular dosing intervals. The aim of this work is to provide model-informed dosing guidance (decision-making aid) for patients, taking into account their characteristics (covariates), as well as clinical or practical constraints (e.g., target concentration for drug efficacy, dosing interval, infusion duration, costs). One important challenge is how to take into account the random components included in population PK models to describe the interindividual (or interoccasion) variability in parameters and drug concentrations (see, e.g., [18]).

Feedback Control for Personalized Medicine
https://doi.org/10.1016/B978-0-32-390171-0.00012-3

Dosing of ABs is frequently based on Monte Carlo simulations from a mathematical PK model to assess the probability of target attainment (PTA) at a population level (see [20]). These probabilities are compared across alternative dosing regimens to select the "optimal" one (see, e.g., [13], [11], and [16]).

In the present chapter, alternative tools are presented, aiming at improving dosing recommendations at the individual level. The pros and cons of these methods are discussed. An input-output analysis leads to a formula which determines the amount of drug to administer (dose), given the covariates (patient level) and the practical/clinical constraints. It can be seen as an open-loop control law. This input-output formula is then used in combination with a "worst-case" analysis to take into account the variability and the unknown realization of the random variables included in the model. The design of an (estimated) state feedback is also considered. The dosing strategies (control laws) reported in this chapter are general methods applicable to any PK system described by a linear time-invariant state-space representation. The numerical results are reported for a case study drug, viz. meropenem, a β-lactam AB used for treating severe infections (see, e.g., [28] and references therein).

3.2 Population pharmacokinetic model

This section describes the state-space representation of a general mammillary compartmental model and its interpretation in terms of PK modeling. A numerical example is given for meropenem.

3.2.1 State-space representation

A population PK (popPK) model describes the PK of a drug of interest at the population level, accounting for different sources of variability in drug concentrations. Mammillary compartmental models are the standard structural models in this field. Such models consist in one central compartment and $n-1$ peripheral compartments. The drug is administrated into either the central compartment or an additional depot compartment. Then, substance exchange takes place between the central and the peripheral compartments (see Fig. 3.1). An n-compartment popPK model with IV administration is described by

$$\dot{x}(t) = Ax(t) + bu(t), \quad y(t) = Cx(t), \tag{3.1}$$

where

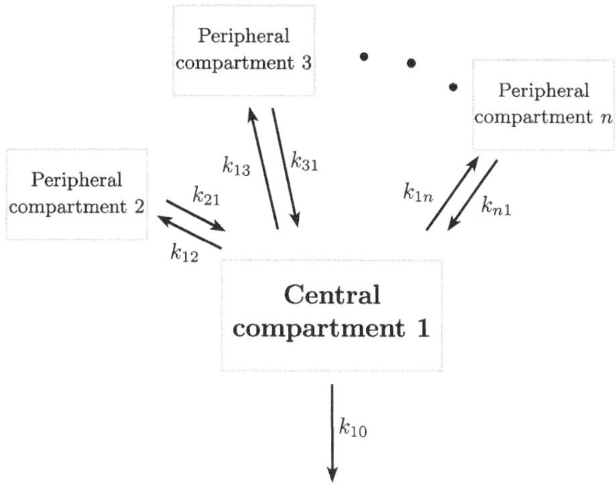

FIGURE 3.1 A general n-compartment mammillary model.

$$A = \begin{pmatrix} -k_{10} - \sum_{j=2}^{n} k_{1j} & k_{21} & k_{31} & \cdots & \cdots & k_{n1} \\ k_{12} & -k_{21} & 0 & 0 & \cdots & 0 \\ k_{13} & 0 & -k_{31} & 0 & \cdots & \vdots \\ \vdots & \vdots & 0 & \ddots & & \vdots \\ \vdots & \vdots & \vdots & \ddots & \ddots & 0 \\ k_{1n} & 0 & 0 & \cdots & 0 & -k_{n1} \end{pmatrix} \tag{3.2}$$

and

$$b = \begin{pmatrix} 1 & 0 & \cdots & 0 \end{pmatrix}^T, \quad C = \begin{pmatrix} 1/S_1 & 0 & 0 & \cdots & 0 \\ 0 & 1/S_2 & 0 & \cdots & 0 \end{pmatrix}. \tag{3.3}$$

The parameters can include random components. In the following deterministic analysis, unless otherwise stated, the PK parameters are assumed to be fixed to their nominal values.

This results in a transfer function of the form $\hat{G}(s) = \begin{pmatrix} \hat{G}_1(s) & \hat{G}_2(s) \end{pmatrix}^T$. The parameters k_{ij} [h^{-1}] represent rate constants and are positive. The state vector $x(t) = \begin{pmatrix} x_1(t) & x_2(t) & x_3(t) & \cdots & x_n(t) \end{pmatrix}^T$ corresponds to the virtual amounts of drug [g] in each compartment. Without lost of generality, x_1 and x_2 correspond to the plasma and the site of infection, respectively. The input function $u(t)$ corresponds to the drug infusion rate [g/h] into the central compartment (plasma). The output vector $y(t) = \begin{pmatrix} y_1(t) & y_2(t) \end{pmatrix}^T$ corresponds to the drug concentrations [mg/L] in the plasma and at the infection site. For all $j \in \{1, 2\}$, S_j is a scaling factor used to convert an amount of drug [g] in concentration in units that are consistent with the observed values (e.g., [mg/L]). It is a scale volume of distribution $S_j = V_j/usv_j$, where usv stands for "unitless scalar value." The *volume of distribution* [L] is a virtual physiological concept defined by the volume in which the drug would be distributed if the compartment drug concentration was equal to the plasma concentration (i.e., for all $i \in \{1, \ldots, n\}$, $c_p[g/L] = x_i/V_i$). Note that the concentrations at the actual effect site are not always described/predicted by the model. This depends on the data available to adequately estimate the values of the relevant parameters. In that case, the matrix C becomes $C = \begin{pmatrix} 1/S_1 & 0 & \cdots & 0 \end{pmatrix}$.

The input function $u(t)$ (see Fig. 3.2) considered here is a right-continuous piecewise constant function that depends on three parameters: the dose [g] denoted by D, the duration of the infusion [h] denoted by Δ, and the time between two doses (dosing interval) [h] denoted by T. For all $t \geq 0$,

$$u(t)\,[\mathrm{g/h}] = \begin{cases} D/\Delta & \text{if } (t \bmod T) < \Delta, \\ 0 & \text{if } (t \bmod T) \geq \Delta, \end{cases} \tag{3.4}$$

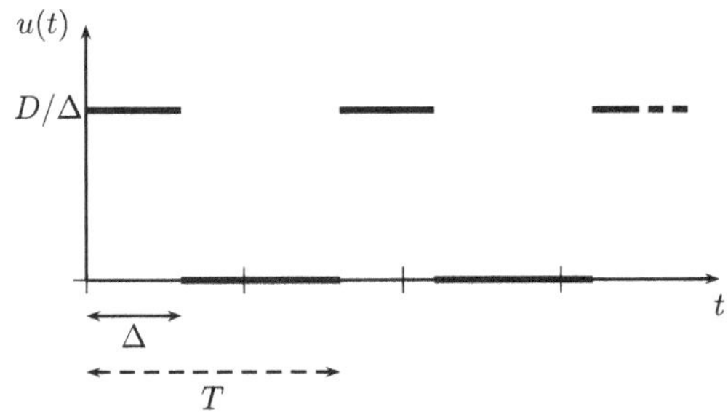

FIGURE 3.2 System input function: drug infusion rate vs. time.

Observe that Δ is necessarily smaller than T. This input structure is consistent with the current mode of administration of meropenem (case study). Administration by IV infusion of a fixed dose (called maintenance dose) at regular intervals is recommended for meropenem (see [17]). Besides, continuous administration for treating severe infection has been studied recently (see, e.g., [28] and [24]). The latter can be seen as a limit case of the mode of administration considered here ($\Delta = T$).

3.2.2 System analysis

Clearly, the matrix $A \in \mathbb{R}^{n \times n}$ is compartmental, i.e., A is Metzler (all its off-diagonal elements are nonnegative) and $\sum_{i=1}^{n} a_{ij} \leq 0$ for all $j \in \{1, \ldots, n\}$ [8, Definition 2.10].

In what follows, for any vector $v \in \mathbb{R}^l$, $v \gg 0$ ($v \ll 0$, respectively) means that all components of v are positive (negative, respectively).

As the state variables of (3.1) represent amounts of drug, it is expected that the state trajectories stay in the nonnegative orthant of $\mathbb{R}^n$ for all nonnegative initial conditions and nonnegative admissible input functions. Since A is a Metzler matrix and b is a nonnegative matrix, it is indeed the case [8, Proposition 4.1].

Proposition 3.1. *The popPK system given by (3.1)–(3.3) is nonnegative.*

Moreover, since the drug is eliminated by the organism, the following result should also be expected.

Proposition 3.2. *The popPK system given by (3.1)–(3.2) is internally (exponentially) stable.*

Proof. Any vector $\mu = \begin{pmatrix} \mu_1 & \mu_2 & \mu_3 & \cdots & \mu_n \end{pmatrix}^T$ such that $\mu_2 = \mu_3 = \cdots = \mu_n > 0$ and $k\mu_2 < \mu_1 < \mu_2$, where

$$k := \frac{k_{12} + k_{13} + \cdots k_{1n}}{k_{10} + k_{12} + k_{13} + \cdots k_{1n}} \in (0, 1),$$

is such that $A^T \mu \ll 0$. The conclusion follows by [4, Proposition 7]. □

3.2.3 Case study: model of meropenem

A dataset of measured concentrations in the plasma and at the site of infection (epithelial lining fluid [ELF]) from patients with severe nosocomial pneumonia was available for model development (PROMESSE study [6]). A two-compartment model provided an adequate fit to the observed plasma concentrations. An additional compartment was added to describe the (sparse) concentrations measured in the ELF:

$$\begin{cases} \dot{x}_1(t) = -k_{10}x_1(t) - k_{12}x_1(t) - k_{13}x_1(t) + \quad k_{21}x_2(t) + k_{31}x_3(t) + u(t), \\ \dot{x}_2(t) = k_{12}x_1(t) - k_{21}x_2(t), \\ \dot{x}_3(t) = k_{13}x_1(t) - k_{31}x_3(t). \end{cases}$$

This system is schematized in Fig. 3.3. In this representation, the physiological parametrization (based on the concepts of volume of distribution and clearances) is preferred. The *(systemic) clearance* (CL) [L/h] is defined by the virtual volume of plasma cleared of drug per time unit, so that $CL = k_{10}V_1$. Likewise, the *intercompartmental clearance* (Q_{1i} and Q_{i1}, $i \in \{2, 3, \ldots, n\}$, [L/h]) is defined by the virtual volume of distribution cleared of drug, to go into the linked compartment, per time unit, so that $Q_{1i} = k_{1i}V_1$ and $Q_{i1} = k_{i1}V_i$. At distribution equilibrium, the same amount of drug goes from the central to the peripheral compartment and from the peripheral to the central compartment, i.e., $k_{1i}x_1 = k_{i1}x_i$, or equivalently $Q_{1i} = Q_{i1} =: Q_i$. To make notations more explicit, V_1, V_2, V_3, Q_2, and Q_3 are written V_c, V_E, V_p, Q_E, and Q_p, respectively (c for central, E for ELF, and p for peripheral), such that

$$CL = k_{10}V_c, \; Q_E = k_{12}V_c, \; Q_E = k_{21}V_E,$$
$$Q_p = k_{13}V_c, \text{ and } Q_p = k_{31}V_p.$$

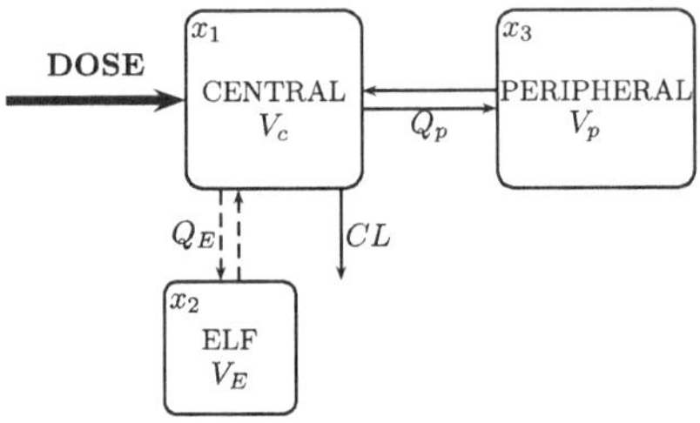

FIGURE 3.3 PopPK model of meropenem describing plasma and ELF concentrations.

Known physiological relationships were incorporated into the model. For example, the renal function influences the total clearance, as shown below with an allometric model:

$$CL_i = TVCL\left(\frac{GFR_i}{GFR_{\text{med}}}\right)^{\theta_{GFR}}, \tag{3.5}$$

where CL_i, which denotes the clearance value of the ith individual (without interindividual variability [IIV]), is a function of the typical parameter value in the population (TVCL) and of the glomerular filtration rate (GFR) of the individual i (GFR_i), normalized by the median value (GFR_{med}). These patient characteristics, which affect the PK of a drug, are called *covariates*.

Random effects are added to the structural model parameters in order to describe the IIV observed in concentrations [2], [21], [22]. The IIV is partly explained by the patient's characteristics (covariates). The remaining variability is said to be *unexplained* and is modeled using random variables. Unexplained IIV is described by exponential models, meaning that parameters are assumed to be log-normally distributed, as exemplified below for the clearance:

$$CL_{\text{IIV}_i} = CL_i \cdot \exp(\eta_i), \tag{3.6}$$

where $CL_{i_{\text{IIV}}}$ is the clearance value of the ith individual (with IIV), CL_i is the *fixed effect* (parameter nominal value of the ith individual), and η_i is the individual realization of the random variable $\eta \sim \mathcal{N}(0, \omega^2)$ (*random effect*). The median of this log-normal distribution is given by CL_i, which corresponds to the realization of the random variable equal to zero.

NONMEM[1] software (Icon Development Solutions, Ellicott City, MD, USA) was used for parameter estimation and model development (including structural model identification and covariate selection). It is a well-known computer program in the PK/PD modeling community, used for more than 30 years to perform population analysis of PK clinical data [2], [14]. Estimated model parameters are reported in Table 3.1. Two covariates were included in the model: the weight (WT) and the GFR, on the volumes of distribution and the clearance, respectively. The unexplained high variability in the distribution of ELF volume may be due to the scarcity of observations and technical factors in the method of measurement (bronchoalveolar lavage [BAL], see [12]).

3.3 Individualized drug dosing

In the scientific literature, model-based AB dosing is mostly based on systemic concentrations, Monte Carlo simulations at the population level, and empirical comparison of successive dosing regimens through PTA analyses (see, e.g., [27]). However, infection-site concentrations are reported to be better predictors of AB effects (see, e.g., [15] and [28]) and there has been an increasing interest in the individualization of AB dosing (see, e.g., [23] and [5]). The advantages of the analytical approaches presented below include the ability to select a dose based on the patient's characteristics (covariates) and on systemic or infection-site concentrations. These methods were successfully applied to meropenem dosing thanks to the numerical popPK model described above and compared to corresponding results of PTA analysis.

3.3.1 Input-output analysis

In the following, the state matrix $A \in \mathbb{R}^{n \times n}$ in (3.1) is assumed to have n distinct eigenvalues. Moreover, observe that the entries of the matrix A correspond to the individual patient, as the PK parameters are computed on the basis of patient characteristics, consistently with the covariate-parameter models, e.g., as in (3.5).

Proposition 3.3. *Consider the popPK system (3.1)–(3.3) with an input function of the form (3.4) where $\Delta \leq T$. The zero-state system response $y(\cdot)$ is given as follows, where N_t denotes the number of administrations already received at time t*

[1]NONMEM stands for NONlinear Mixed Effects Modeling; computer program implemented in Fortran90/95, originally developed by Lewis Sheiner and Stuart Beal at the University of California [2].

(including the ongoing administration, if appropriate): for all $j \in \{1, 2\}$ *and for all* $t \geq 0$,

- *if* $(t \bmod T) < \Delta$ *(during infusion),*

$$y_j(t) = \frac{D}{\Delta}\left[\underbrace{\sum_{l=0}^{N_t-2}\sum_{i=1}^{n}\frac{F_{ji}}{\lambda_i}e^{\lambda_i(t-lT)}(1-e^{-\lambda_i\Delta})}_{\text{previous administrations}} + \underbrace{\sum_{i=1}^{n}\frac{F_{ji}}{\lambda_i}(e^{\lambda_i(t-(N_t-1)T)}-1)}_{\text{ongoing infusion}}\right], \tag{3.7}$$

- *if* $(t \bmod T) \geq \Delta$ *(after infusion),*

$$y_j(t) = \frac{D}{\Delta}\sum_{l=0}^{N_t-1}\sum_{i=1}^{n}\frac{F_{ji}}{\lambda_i}e^{\lambda_i(t-lT)}(1-e^{-\lambda_i\Delta}). \tag{3.8}$$

TABLE 3.1 PopPK model of meropenem: parameter estimates.

Parameter	Estimate[a]	%RSE[b]	bootstrap 95% CI[c]
CL (L/h)	7.94	4.95	7.34–8.54
GFR effect on CL [θ_{GFR}][d]	0.722	8.82	0.581–0.816
IVV on CL [ω^2 (%CV)]	0.126 (35.5)	18.8	0.0817–0.157
$\boldsymbol{V_c}$ (L)	13.6	6.26	11.5–15.4
WT effect on V_C [θ_{WT}][e]	0.949	30.1	0.609–1.68
IVV on V_C [ω^2 (%CV)]	0.140 (37.4)	33.6	0.0827–0.235
$\boldsymbol{Q_E}$ (L/h)	6.73	18.3	4.02–14.4
$\boldsymbol{V_E}$ (L)	4.08	33.3	1.17–5.87
WT effect on V_E [θ_{WT}][e]	1.04	114	−5.39–3.02
IVV on V_E [ω^2 (%CV)]	1.76 (133)	41.1	0.909–4.89
$\boldsymbol{Q_p}$ (L/h)	8.22	18.3	5.70–11.5
IVV on Q_P [ω^2 (%CV)]	0.187 (43.2)	56.1	0.0206–0.349
$\boldsymbol{V_p}$ (L)	10.1	15.5	7.75–13.5
$\boldsymbol{usv_E}$ (V_E/S_E)	249	9.68	233–324
Eprop plasma conc. [σ_p^2 (%CV)]	0.0240 (15.5)	11.8	0.0170–0.0292
Eadd plasma conc. [σ_a^2 (mg/L)]	0.208 (0.456)	32.5	0.114–0.379
Eprop ELF conc. [σ_p^2 (%CV)]	0.404 (63.6)	21.3	0.254–0.547

CL, clearance; GFR, glomerular filtration rate; V_C*, volume of distribution of the central compartment; WT, body weight;* Q_E*, intercompartmental clearance between the central and ELF compartments;* V_E*, volume of distribution of the ELF compartment;* Q_P*, intercompartmental clearance between the central and peripheral compartments;* V_P*, volume of distribution of the peripheral compartment; IIV, interindividual variability;* ω^2*, variance of the related random variable; usv; unitless scalar value; S, scaling factor; Eprop, proportional error; Eadd, additional error;* $\sigma^2_{p,a}$*, variances of the related random variables.*

[a] *Estimates provided by NONMEM.*

[b] *Relative standard error of estimates.*

[c] *Confidence interval based on the 2.5th and 97.5th percentiles of the bootstrapped distributions generated from 1000 resampled datasets.*

[d] *Allometric covariate model, normalized to a GFR level of 65 mL/min.*

[e] *Allometric covariate model, normalized to a body weight of 75 kg.*

Coefficients F_{ji} and λ_i ($i \in \{1, 2, 3, \ldots n\}$, $j \in \{1, 2\}$) are directly related to the model parameters: λ_i denote the eigenvalues of the matrix A and F_{ji} are the residuals of the transfer function $\hat{G}_j(s)$ in the eigenvalues λ_i, i.e., $F_{ji} = \lim_{s \to \lambda_i} \left(\hat{G}_j(s)\,(s - \lambda_i)\right)$.

Proof.

Preliminary remark. The *Laplace transform* of a function $f : \mathbb{R} \to \mathbb{C}$ with support on $\mathbb{R}^+$ is the function $\hat{f}$ defined by

$$\hat{f}(s) = \mathcal{L}[f(\cdot)](s) = \int_0^\infty e^{-st} f(t)\, dt$$

for all $s \in \mathbb{C}$ such that $\mathrm{Re}[s] > \sigma_0$, where σ_0, called the absolute convergence abscissa, defines an existence right-half plane.

A delay of T units in the time domain corresponds to a multiplication by e^{-sT} in the Laplace domain, i.e.,

$$\mathcal{L}^{-1}[e^{-\cdot T} \hat{f}(\cdot)](t) = f(t - T), \quad t \geq T.$$

Step 1: transfer function. Let $L(s)$ denote the matrix $(sI - A)$, whose characteristic polynomial is given by

$$det(L(s)) = (s - \lambda_1)(s - \lambda_2)(s - \lambda_3) \cdots (s - \lambda_n).$$

Each component $\hat{g}_j$ ($j \in \{1, \ldots, p\}$) of the transfer function $\hat{G}(s)$ can be written in partial fraction expansion as

$$\hat{g}_j(s) = \frac{F_{j1}}{s - \lambda_1} + \frac{F_{j2}}{s - \lambda_2} + \frac{F_{j3}}{s - \lambda_3} + \cdots + \frac{F_{jn}}{s - \lambda_n},$$

where $F_{j1}, F_{j2}, F_{j3}, \ldots F_{jn}$ are the residuals of $\hat{g}_j(s)$ in $\lambda_1, \lambda_2, \lambda_3, \ldots \lambda_n$, respectively.

Step 2: Laplace transform of the input function. The input function can be written as a superposition of functions of the form

$$U_i(t) = \frac{D}{\Delta} \cdot \mathbf{1}_{[iT, iT+\Delta[}(t), \quad t \geq 0,$$

where $i \in \mathbb{N}$ and $\mathbf{1}_I$ is the characteristic function of the subset I. Let $N \in \mathbb{N}_0$ be the number of administrations. The input function $u(t)$ is written as

$$u(t) = \sum_{i=0}^{N-1} U_i(t), \quad t \geq 0.$$

The reference input function, corresponding to the first administration, is denoted $U(t)$ and given by

$$U(t) = U_0(t) = \frac{D}{\Delta} \cdot \mathbf{1}_{[0,\Delta[}(t), \quad t \geq 0.$$

The Laplace transform of the latter function is given by

$$\hat{U}(s) = \frac{D}{\Delta}\left(\frac{1}{s} - \frac{e^{-\Delta s}}{s}\right).$$

The other functions $U_i(t)$ ($i \in \{1, \ldots, N-1\}$) are equal to the reference input with a delay of iT time units, i.e.,

$$U_i(t) = U(t - iT), \quad t \geq iT.$$

According to the preliminary remark, they are expressed in the Laplace domain by

$$\hat{U}_i(s) = e^{-siT}\hat{U}(s).$$

Using the linearity of the Laplace transformation, we finally obtain

$$\hat{u}(s) = \sum_{i=0}^{N-1} e^{-siT} \frac{D}{\Delta}\left(\frac{1}{s} - \frac{e^{-\Delta s}}{s}\right).$$

Step 3: Time domain response. In the Laplace domain, the zero-state system response is

$$\hat{y}(s) = \hat{G}(s)\hat{u}(s).$$

For $j \in \{1, \ldots p\}$,

$$\hat{y}_j(s) = \hat{g}_j(s)\hat{u}(s) = \frac{D}{\Delta} \cdot \sum_{l=0}^{N-1} \underbrace{e^{-slT}}_{delay} \hat{Y}_j(s), \tag{3.9}$$

where

$$\hat{Y}_j(s) = \sum_{i=1}^{n} \frac{F_{ji}}{s - \lambda_i}\left(\frac{1}{s} - \frac{e^{-\Delta s}}{s}\right) = \hat{h}_j(s) - \underbrace{e^{-\Delta s}}_{delay} \hat{h}_j(s), \tag{3.10}$$

where $\hat{h}_j(s) = \sum_{i=1}^{n} \frac{F_{ji}}{s - \lambda_i}\frac{1}{s}$ (proper rational function). Its partial fraction decomposition (where $\lambda_{n+1} = 0$) is given by

$$\hat{h}_j(s) = \frac{\widetilde{F}_{j1}}{s - \lambda_1} + \frac{\widetilde{F}_{j2}}{s - \lambda_2} + \frac{\widetilde{F}_{j3}}{s - \lambda_3} + \cdots \frac{\widetilde{F}_{jn}}{s - \lambda_n} + \frac{\widetilde{F}_{j(n+1)}}{s - \lambda_{n+1}},$$

where $\widetilde{F}_{j1}, \widetilde{F}_{j2}, \widetilde{F}_{j3}, \dots \widetilde{F}_{j(n+1)}$ are the residuals of $\hat{h}_j(s)$ in $\lambda_1, \lambda_2, \lambda_3, \dots \lambda_{n+1}$, respectively. We can easily show that, for all $i \in \{1, 2, 3, \dots n\}$,

$$\widetilde{F}_{ji} = \frac{F_{ji}}{\lambda_i} \quad \text{and} \quad \widetilde{F}_{j(n+1)} = -\left(\frac{F_{j1}}{\lambda_1} + \frac{F_{j2}}{\lambda_2} + \frac{F_{j3}}{\lambda_3} + \cdots \frac{F_{jn}}{\lambda_n}\right).$$

Laplace transform formulas enable us to write the inverse Laplace transform of $\hat{h}_j(s)$ as follows:

$$h_j(t) = \mathcal{L}^{-1}[\hat{h}_j(\cdot)](t) = \sum_{i=1}^{n+1} \tilde{F}_{ji} e^{\lambda_i t} = \sum_{i=1}^{n} \frac{F_{ji}}{\lambda_i}(e^{\lambda_i t} - 1), \quad t \geq 0.$$

Using the preliminary remark, we get

$$\begin{aligned} Y_j(t) &= \mathcal{L}^{-1}[\hat{Y}_j(\cdot)](t) \\ &= h_j(t) - h_j(t - \Delta), \quad t \geq \Delta \\ &= h_j(t) - h_j(t - \Delta)\mathbf{1}_{[\Delta, +\infty[}(t), \quad t \geq 0. \end{aligned}$$

Similarly, we can deduce from (3.9) that, for all $t \geq 0$,

$$\begin{aligned} y_j(t) &= \mathcal{L}^{-1}[\hat{y}_j(\cdot)](t) \\ &= \frac{D}{\Delta} \sum_{l=0}^{N-1} Y_j(t - lT)\mathbf{1}_{[lT+\infty[}(t), \quad t \geq 0 \\ &= \frac{D}{\Delta} \sum_{l=0}^{N-1} \left[h_j(t - lT)\mathbf{1}_{[lT, +\infty[}(t) - h_j(t - lT - \Delta)\mathbf{1}_{[lT+\Delta, +\infty[}(t)\right], \end{aligned}$$

where $h_j(t) = \sum_{i=1}^{n} \frac{F_{ji}}{\lambda_i}(e^{\lambda_i t} - 1)$; hence, identities (3.7) and (3.8) hold. □

Let us denote by $y^N(\tilde{t})$ ($\tilde{t} \in [0, T[$) the system response on the Nth dosing interval, i.e., $y(t) = y^N(\tilde{t})$, where $t \in [(N-1)T, NT[$ and $\tilde{t} = t \bmod T$. The new variable $\tilde{t}$ indicates the position in the dosing interval after N administrations. Due to system stability, the concentration trajectory converges to an equilibrium trajectory, which will be called steady output (known in the pharmacology literature as the "steady state" or "plateau"), i.e.,

$$\lim_{N \to \infty} y^N(\tilde{t}) = y^\infty(\tilde{t}).$$

Proposition 3.4. *Consider the popPK system (3.1)–(3.3) with an input function of the form (3.4) where $\Delta \leq T$. The output trajectory converges to an equilibrium* *trajectory, i.e., for all $\tilde{t} \in [0, T[$,*

$$y^N(\tilde{t}) - y^\infty(\tilde{t}) \xrightarrow{N\to\infty} 0,$$

where the plateau y^∞ is given, for all $j \in \{1, 2\}$, by:

- *if $\tilde{t} \in [0, \Delta]$,*

$$y_j^\infty(\tilde{t}) = \frac{D}{\Delta}\left[\sum_{i=1}^{n} \frac{F_{ji}}{\lambda_i} \frac{e^{\lambda_i \tilde{t}}(1 - e^{-\lambda_i \Delta})}{e^{-\lambda_i T} - 1} + \sum_{i=1}^{n} \frac{F_{ji}}{\lambda_i}(e^{\lambda_i \tilde{t}} - 1)\right],$$

- *if $\tilde{t} \in [\Delta, T[$,*

$$y_j^\infty(\tilde{t}) = \frac{D}{\Delta}\sum_{i=1}^{n} \frac{F_{ji}}{\lambda_i} \frac{e^{\lambda_i(\tilde{t}-T)}(1 - e^{-\lambda_i \Delta})}{e^{-\lambda_i T} - 1}.$$

Proof. Let $j \in \{1, \ldots, p\}$ and $t \geq 0$ such that $(t \bmod T) \geq \Delta$. Let $N \in \mathbb{N}_0$ be the number of administrations received at time t. The zero-state system response $y_j(t)$ (see (3.8)) is given by

$$y_j(t) = \frac{D}{\Delta}\sum_{i=1}^{n} \frac{F_{ji}}{\lambda_i}(1 - e^{-\lambda_i \Delta})e^{\lambda_i t}\sum_{l=0}^{N-1} e^{-\lambda_i l T}$$

or, equivalently,

$$y_j(t) = \frac{D}{\Delta}\sum_{i=1}^{n} \frac{F_{ji}}{\lambda_i}(1 - e^{-\lambda_i \Delta})e^{\lambda_i t}\frac{1 - e^{-\lambda_i N T}}{1 - e^{-\lambda_i T}}.$$

We set $t = \tilde{t} + (N-1)T$ (i.e., $\tilde{t} = t \bmod T$), hence $\Delta \leq \tilde{t} < T$ indicates the position on the dosing interval $[0, T[$ after N administrations. Consequently, $y_j(t)$ can be written

$$y_j^N(\tilde{t}) = \frac{D}{\Delta}\sum_{i=1}^{n} \frac{F_{ji}}{\lambda_i}(1 - e^{-\lambda_i \Delta})\frac{e^{\lambda_i(N-1)T} - e^{-\lambda_i T}}{1 - e^{-\lambda_i T}}e^{\lambda_i \tilde{t}}.$$

Since, for all $i \in \{1, \ldots, n\}$, $\mathrm{Re}(\lambda_i) < 0$ (system stability), the asymptotic response reads

$$y_j^\infty(\tilde{t}) := \lim_{N\to\infty} y_j^N(\tilde{t}) = \frac{D}{\Delta}\sum_{i=1}^{n} \frac{F_{ji}}{\lambda_i}(1 - e^{-\lambda_i \Delta})\frac{e^{-\lambda_i T}}{e^{-\lambda_i T} - 1}e^{\lambda_i \tilde{t}}.$$

A similar development can be applied for $t \geq 0$ such that $(t \bmod T) < \Delta$ (or, equivalently, $0 \leq \tilde{t} < \Delta$). Let $N \in \mathbb{N}_0$ be the number of administrations received (including the ongoing administration) at time t. The zero-state system response $y_j(t)$ (see (3.7)) is given by

$$y_j(t) = \frac{D}{\Delta}\left[\sum_{i=1}^{n} \frac{F_{ji}}{\lambda_i}(1 - e^{-\lambda_i \Delta})e^{\lambda_i t} \sum_{l=0}^{N-2} e^{-\lambda_i l T} + \sum_{i=1}^{n} \frac{F_{ji}}{\lambda_i}(e^{\lambda_i (t-(N-1)T)} - 1)\right],$$

i.e.,

$$y_j(t) = \frac{D}{\Delta}\left[\sum_{i=1}^{n} \frac{F_{ji}}{\lambda_i}(1 - e^{-\lambda_i \Delta})e^{\lambda_i t} \frac{1 - e^{-\lambda_i (N-1)T}}{1 - e^{-\lambda_i T}} + \sum_{i=1}^{n} \frac{F_{ji}}{\lambda_i}(e^{\lambda_i (t-(N-1)T)} - 1)\right].$$

Hence, for $\tilde{t} \in [0, \Delta[$,

$$y_j^N(\tilde{t}) = \frac{D}{\Delta}\left[\sum_{i=1}^{n} \frac{F_{ji}}{\lambda_i}(1 - e^{-\lambda_i \Delta})\frac{e^{\lambda_i (N-1)T} - 1}{1 - e^{-\lambda_i T}} e^{\lambda_i \tilde{t}} + \sum_{i=1}^{n} \frac{F_{ji}}{\lambda_i}(e^{\lambda_i \tilde{t}} - 1)\right].$$

Since, for all $i \in \{1, \ldots, n\}$, $\mathrm{Re}(\lambda_i) < 0$ (system stability), the asymptotic response reads

$$\begin{aligned} y_j^{\infty}(\tilde{t}) &:= \lim_{N \to \infty} y_j^N(\tilde{t}) \\ &= \frac{D}{\Delta}\left[\sum_{i=1}^{n} \frac{F_{ji}}{\lambda_i}(1 - e^{-\lambda_i \Delta})\frac{1}{e^{-\lambda_i T} - 1} e^{\lambda_i \tilde{t}} + \sum_{i=1}^{n} \frac{F_{ji}}{\lambda_i}(e^{\lambda_i \tilde{t}} - 1)\right]. \quad \square \end{aligned}$$

It follows from Proposition 3.4 that, at steady state, the concentrations at the end and beginning of the dosing interval are equal, $\lim_{\tilde{t} \underset{>}{\to} T} y^{\infty}(\tilde{t}) = y^{\infty}(0)$, such that we can define

$$y^{\infty}(T) := y^{\infty}(0).$$

The concentration $y_1^{\infty}(T)$ corresponds to the minimal *plasma* concentration on the dosing interval (the lowest concentration [c_{trough}] observed just before a new administration). However, if there is a time disconnect between systemic and infection-site concentrations (due to, e.g., delayed/slow tissue penetration), $y_2^{\infty}(T)$ does not correspond to the minimal concentration at the site of infection.

In the following, y and y^{∞} are denoted by $y(D, \Delta, T; t)$ and $y^{\infty}(D, \Delta, T; \tilde{t})$, respectively, in order to highlight the dependence with respect to the input parameters. From now on, the infusion duration Δ [h] and the length of the dosing interval T [h] are assumed to be fixed.

3.3.2 Input-output formula for drug dosing

A natural approach consists in solving the following equation with respect to D:

$$y_j^{\infty}(D, \Delta, T; \tilde{t} = t^*) = \alpha \tag{3.11}$$

in order to compute the dose D [g] (*maintenance dose*) required to reach, at steady state, a target (systemic or infection-site) concentration α [mg/L] at a given time $t^* \in [0, T]$. This relevant time t^* depends on the clinical objective. If it is required to reach this target concentration at the very first infusion, we should solve the following equation with respect to D:

$$y_j(D, \Delta, T; t = t^*) = \alpha.$$

This dose is called *loading dose* and is denoted by D_L. The loading dose is thus given at the first administration, i.e., for all $t \in [0, \Delta[$, $u(t) = D_L/\Delta$. The maintenance dose is then given at the following administrations, i.e., for all $t \in [iT, iT + \Delta[$ $(i \in \mathbb{N}_0)$, $u(t) = D/\Delta$.

Meropenem is a time-dependent AB, meaning that treatment efficacy is measured in terms of percentage of time between two doses during which the drug concentration exceeds the minimal inhibitory concentration (MIC) (%T>MIC). In critically ill patients, it can be required to achieve 100%T>MIC to ensure optimal clinical outcome (see, e.g., [7] and [23]). Clinical breakpoints for MIC are computed from in vitro experiments and provided by EUCAST (the European Committee on Antimicrobial Susceptibility Testing). They are also reported in different summaries of product characteristics of medicinal products containing meropenem (e.g., [17]). Meropenem has a slight time disconnect between systemic and infection-site concentrations, such that the time t^* related to 100%T>MIC at the site of infection (ELF) is T. In this case, the analytical solution to (3.11) is given in the proposition below.

Proposition 3.5. *Consider the popPK system (3.1)–(3.3) with an input function of the form (3.4). For any target concentration $\alpha > 0$ [mg/L], the maintenance dose [g] required to maintain the steady output trajectory y_j^{∞} above the lower bound α is given by*

$$D = \frac{\alpha \Delta}{\displaystyle\sum_{i=1}^{n} \frac{F_{ji}}{\lambda_i} \frac{1 - e^{-\lambda_i \Delta}}{e^{-\lambda_i T} - 1}}, \tag{3.12}$$

where j denotes the regulated output ($j = 1$ for plasma concentration and $j = 2$ for infection-site concentration). For all $i \in \{1, 2, 3, \ldots n\}$, λ_i denote the eigenvalues of the matrix A and F_{ji} are the residuals of the transfer function $\hat{G}_j(s)$ in λ_i.

Formula (3.12) is called input-output (I/O) formula in what follows. The block diagram in Fig. 3.4 represents the resulting open-loop system.

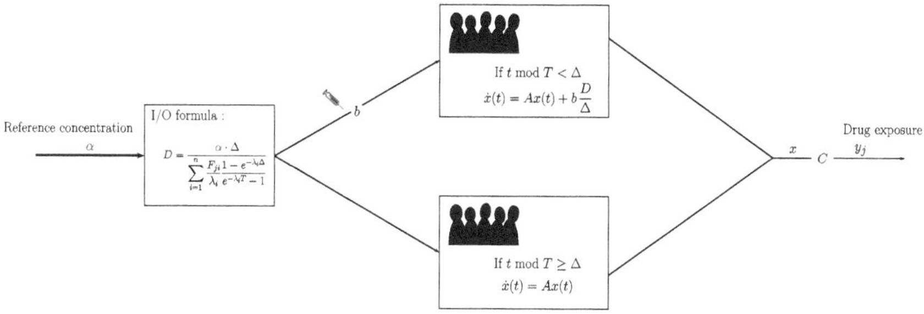

FIGURE 3.4 Block diagram representing the open-loop system. $[A, b, C]$, model specific to the individual patient; α, target (minimal) concentration; y_j, system output (concentration in the plasma, $j = 1$; at the site of infection, $j = 2$). The coefficients F_{ji} and λ_i ($i \in \{1, 2, 3, \ldots n\}$) are directly related to the model parameters.

The parameters F_{ij} and λ_i are directly related to the PK parameters and, consequently, to the covariate values and the realization of the random variables. However, these realizations are, a priori, unknown. A way to overcome this difficulty is to incorporate the I/O formula into a worst-case analysis.

3.3.3 "Worst-case" analysis

A sensitivity analysis was performed on the popPK model described in Section 3.2.3 and revealed that, among all the PK parameters, the clearance has the most important influence on the model predictions. Therefore, we call "worst case" the case of a patient for which the clearance value corresponds to the worst clearance regarding the objective to be achieved. All the other PK parameters are assumed to be fixed at their nominal/median value (realizations of the related random variables are set to zero).

Proposition 3.6. *Consider the popPK system (3.1)–(3.2), where $b \in \mathbb{R}^{n\times 1}$ is any nonnegative matrix. The state trajectory $x(t; CL)$ is decreasing with respect to the clearance parameter, i.e., if $0 < CL_1 \leq CL_2$ and $x(0, CL_1) \geq x(0, CL_2)$, then, for all $t \geq 0$, $x(t; CL_1) \geq x(t; CL_2)$.*

Proof. Using the physiological parametrization, the state matrix A reads as

$$A = \begin{pmatrix} -\frac{CL}{V_1} - \sum_{j=2}^{n} \frac{Q_j}{V_1} & \frac{Q_2}{V_2} & \frac{Q_3}{V_3} & \cdots \\ \frac{Q_2}{V_1} & -\frac{Q_2}{V_2} & 0 & \cdots \\ \frac{Q_3}{V_1} & 0 & -\frac{Q_3}{V_3} & 0 \\ \vdots & \vdots & 0 & \ddots \end{pmatrix}.$$

In the following, A is denoted A_{CL} to highlight the dependence with respect to the clearance parameter. The state trajectories $x(t; CL_1)$ and $x(t; CL_2)$ (where $CL_1 \leq CL_2$) are the solutions of the following Cauchy problems, respectively:

$$\begin{cases} \dot{x} = A_{CL_1} x + bu, & x(0) = x_0, \\ \dot{\tilde{x}} = A_{CL_2} \tilde{x} + bu, & \tilde{x}(0) = \tilde{x}_0, \end{cases}$$

where $0 \leq \tilde{x}_0 \leq x_0$. The state matrix A_{CL_2} can be written as

$$A_{CL_2} = A_{CL_1} + \begin{pmatrix} -\epsilon & 0 & \cdots \\ 0 & 0 & \cdots \\ \vdots & \vdots & \ddots \end{pmatrix},$$

where $\epsilon = \frac{CL_2 - CL_1}{V_1}$. The dynamics of $x - \tilde{x}$ is described by the following differential equation:

$$\frac{d}{dt}(x - \tilde{x}) = A_{CL_1}(x - \tilde{x}) + (A_{CL_1} - A_{CL_2})\tilde{x}, \tag{3.13}$$

which obviously describes a nonnegative system. Consequently, $x(t; CL_1) \geq x(t; CL_2)$ for all $t \geq 0$. □

Therefore, this result holds also for the equilibrium trajectory, i.e., whenever $0 < CL_1 \leq CL_2$, for all $\tilde{t} \in [0, T]$, $y^{\infty}(\tilde{t}; CL_1) \geq y^{\infty}(\tilde{t}; CL_2)$. Thanks to this result and given that the objective to be achieved is to maintain the steady output above a fixed level (MIC) for 100% of the time of the dosing interval, the worst clearance corresponds to its highest value. However, according to identity (3.6), i.e.,

$$CL_{\mathrm{IIV}_i} = CL_i \cdot \exp(\eta_i),$$

we define the *"worst-case" realization* at $p \cdot 100\%$, where $0 < p < 1$ and $p = k/q$, as the kth q-quantile. For example, the "worst-case" realization at 90% ($p = 0.9$) is the 9th decile: $\eta^* = 1.28 \cdot \omega$.

Proposition 3.7. *Consider the popPK system (3.1)–(3.3) with an input function of the form (3.4). Assume that the clearance parameter is set to its "worst-case" value at $p \cdot 100\%$. Then, for any target concentration level $\alpha > 0$ [mg/L], the maintenance dose [g] computed according to the I/O formula (3.12) enables to maintain the steady output trajectory y_j^{∞} above the threshold α for $p \cdot 100\%$ of the simulated individuals, provided that all their PK parameters, except the clearance, are fixed to their nominal value.*

3.3.4 PTA analysis

The methods described in Sections 3.3.2 and 3.3.3 to compute an individualized dose were applied to the developed model of meropenem (see Section 3.2.3). Fig. 3.5 represents a simulated PK profile y_j^∞ (here, $j = 2$) using this model and an arbitrary dose. If we consider a high number of simulations denoted by n_{sim}, the probability of reaching 100%T>MIC is estimated by

$$\frac{\#\{y_j^\infty \mid y_j^\infty(T) > MIC\}}{n_{\text{sim}}}.$$

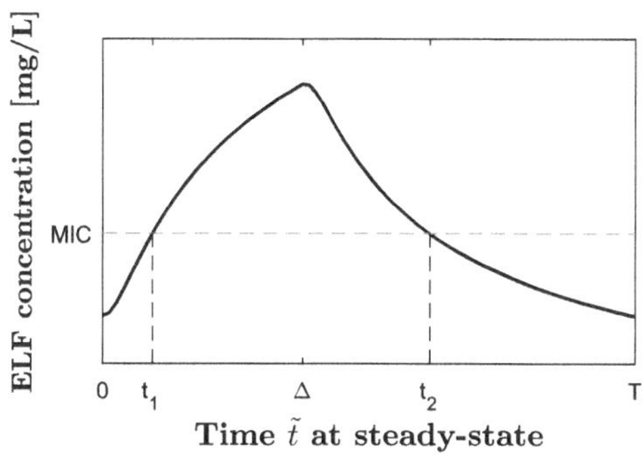

FIGURE 3.5 Simulated PK profile.

For each simulated PK profile, t_1 and t_2 denote the time points such that $t_1 < t_2$, $y_j^\infty(t_1) = \text{MIC}$, and $y_j^\infty(t_2) = \text{MIC}$. The probability of reaching 50%T>MIC is estimated by

$$\frac{\#\{y_j^\infty \mid t_2 - t_1 \geq T/2\}}{n_{\text{sim}}}.$$

An analysis of the PTA can be used to compare different dosing regimens. Two doses were computed by the I/O formula ($\alpha = 2$ mg/L, $j = 2$, $T = 8$ h and $\Delta = 0.5$ h), in the nominal case and in the "worst case" at 90% (see Propositions 3.5 and 3.7, respectively). The respective dosing regimens (1.52 g/8 h, 30′ infusion; 4.34 g/8 h, 30′ infusion) were simulated using 1000 virtual patients with a body weight of 75 kg and a GFR of 65 mL/min (Monte Carlo simulations). Fig. 3.6 represents, for each dosing regimen, the percentage of simulated subjects which reached the PK/PD targets (100%T>MIC or 50%T>MIC) for different MICs in ELF (site of infection). They are compared with the dose recommended by the provider for adult patients with pneumonia: 1 g by IV infusion over 30 minutes every 8 hours [17]. An MIC of 2 mg/L is the EUCAST MIC breakpoint for susceptible strains.

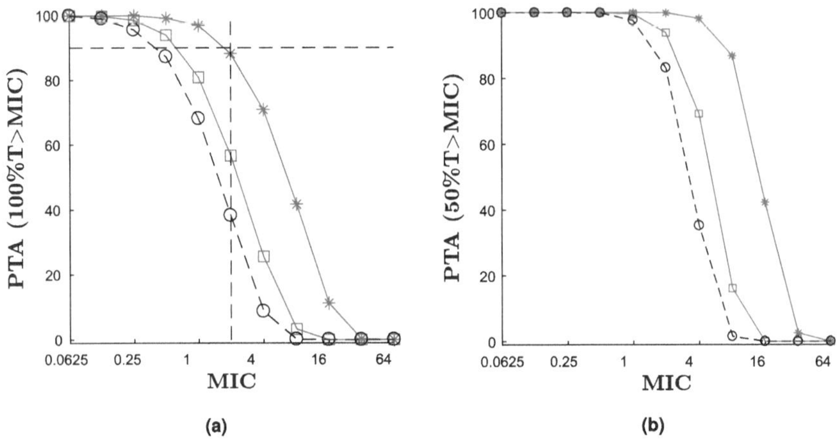

FIGURE 3.6 PTA in ELF based on model predictions ($n_{\text{sim}} = 1000$). The blue line with squares, red line with stars, and black line with circles correspond to the dosing regimens associated to the nominal case, the "worst case," and the provider recommendation, respectively.

The dose associated to the "worst case" provides a probability of success of about 90% for an MIC of 2 mg/L, which was the target concentration used in the I/O formula. This conservative dose (4.34 g) is obviously higher than the first one (1.52 g). However, this dosing strategy enables to ensure that a larger proportion of the patients will reach the PK/PD target (Fig. 3.6). That leaves open the question of a trade-off between the dose (in terms of risk of toxicity [3], cost, etc.) and the probability to achieve the target.

3.3.5 Drawbacks and advantages of the I/O approach

The main drawbacks are related to its open-loop nature and the deterministic approach of a problem including variability. However, this method leads to a feedback control law that is developed in the following section. An additional weakness is the determination of the relevant time t^* (see Eq. (3.11)) in the case of AB with important time disconnect between systemic and local concentrations. Besides, this I/O formula for dose selection has several clinical advantages:

- It is an analytical method, in contrast to the empirical (simulation-based) PTA approach.
- Provided that the model describes the local concentrations, the I/O formula can be used with a target concentration at the site of effect/infection ($j = 2$).
- The covariates are properly taken into consideration to compute an individualized drug dosing (modulo the realization of the random vari-

ables), unlike the standard PTA method, which is most commonly performed at a population level.

3.4 State estimation and feedback control law

In this section, we distinguish the observed output (y_1) and the regulated output (y_1 or y_2) using distinct output matrices:

$$C_{\text{mes}} = \begin{pmatrix} 1/S_1 & 0 & 0 & \cdots & 0 \end{pmatrix} \in \mathbb{R}^{1\times n}$$

and

$$C_{\text{reg}} = \begin{pmatrix} 1/S_1 & 0 & 0 & \cdots & 0 \end{pmatrix} \quad \text{or} \quad C_{\text{reg}} = \begin{pmatrix} 0 & 1/S_2 & 0 & \cdots & 0 \end{pmatrix},$$

where, for all $j \in \{1, 2\}$, S_j converts the state component x_j in concentration both in plasma ($j = 1$) and at the site of infection ($j = 2$).

Since the plasma concentration is *not* continually measured, we need to transform the continuous-time dynamical system (3.1) in a discrete-time setting in order to develop a state estimator. Let us denote by N_1 and N_2 the numbers of sampling intervals on any interval of the form $[iT, iT + \Delta[$ and $[iT + \Delta, (i + 1)T[$, respectively ($i \in \mathbb{N}$). The parameters N_1 and N_2 are chosen to be the smallest integer such that

$$\frac{T}{\Delta} = 1 + \frac{N_2}{N_1}$$

(i.e., same discretization steps). In the following, we will use the notations $h := \Delta/N_1$ and $N := N_1 + N_2$ to denote the discretization step and the total number of sampling intervals on any interval of the form $[iT, (i + 1)T[$, respectively ($Nh = T$). The recurrence equation describing the state dynamics in discrete time is given, for all $k \in \mathbb{N}$, by

$$x[k+1] = A_d x[k] + b_d u[k], \quad y[k] = C_{\text{mes}} x[k], \tag{3.14}$$

where $A_d = e^{Ah}$, $b_d = \int_0^h e^{As} b\, ds$, and

$$u[k] = \begin{cases} D_i/\Delta & \text{if } (k \bmod N) < N_1 \quad (i = \lfloor k/N \rfloor), \\ 0 & \text{if } (k \bmod N) \geq N_1, \end{cases}$$

where D_i denotes the dose administered on the $(i + 1)$th dosing interval and the symbol $\lfloor\cdot\rfloor$ denotes the integer part of a real number.

The discrete-time dynamical system above is obtained by discretizing the integral form of the state trajectory of (3.1):

$$x(t) = e^{A(t-t_0)} x(t_0) + \int_{t_0}^{t} e^{A(t-\tau)} b u(\tau) d\tau,$$

where the control input $u(t)$ is assumed constant on the sampling intervals. Moreover, since $u(t)$ is assumed to be piecewise constant on the dosing intervals, for the same initial condition ($x[0] = x(0)$), the solution of (3.14) is equal to the solution of the continuous-time model at the sampling times, i.e., for all $k \in \mathbb{N}$, $x[k] = x(kh)$.

The nonnegativity and the internal stability of the discrete-time model [8, Chapter 2] are directly derived from the nonnegativity and the internal stability of the continuous-time system.

The time scale of the state estimator is larger than the one of the dynamics (3.14) because we assume that the plasma concentration is measured only before each new administration, i.e., at time iT for all $i \in \mathbb{N}_0$. The state estimator is described, for all $i \in \mathbb{N}$, by

$$\begin{cases} \tilde{x}[i+1] = \mathcal{A}\tilde{x}[i] + \mathcal{B}u[i] + \mathcal{L}(\tilde{y}[i] - y[iN]), \quad \tilde{x}[0] = 0, \\ \tilde{y}[i] = C_{\text{mes}}\tilde{x}[i], \end{cases}$$

where, $\mathcal{A} = A_d^N$, $\mathcal{B} = \sum_{l=N_2}^{N-1} A_d^l b_d$. For all $i \in \mathbb{N}$, $\tilde{x}[i]$ is the estimated state of $x[iN]$ and, therefore, of $x(iT)$. The output injection matrix $\mathcal{L}$ has to be designed such that $\mathcal{A} + \mathcal{L}C$ is stable, i.e., for all $\lambda \in \sigma(\mathcal{A} + \mathcal{L}C)$, $|\lambda| < 1$. Standard methods, such as pole placement or Kalman filter, can be used to determine $\mathcal{L}$.

The estimated state feedback described below is a heuristic method which is reported as an algorithm. It is based on the superposition principle: for any doses $D_1, D_2 > 0$, $y^\infty(D_1 + D_2, \Delta, T; \tilde{t}) = y^\infty(D_1, \Delta, T; \tilde{t}) + y^\infty(D_2, \Delta, T; \tilde{t})$. In this approach, we assume that empirical Bayes estimations (EBEs) are performed at a certain moment to recalculate the individual's PK parameter after appropriate measurements of concentrations, as implemented in different software and websites (see, e.g., [30]). The updated model describes then how the drug is behaving in that particular patient [29]. The Bayesian estimations of the random parameters are the most likely values given the observations and the population model, i.e., $\eta_i^* = \arg\max P(\eta_i | y_i, A, b, C)$ (see, e.g., [19] and [21]).

Algorithm 3.1 (Estimated state feedback).
Data: The mode of administration (T and Δ); the target concentration level (*reference* α) of the regulated output; the typical parameter estimates and the individual's covariate values ($\rightarrow$ nominal or "worst-case" system $[A, b, C_{\text{mes}}/C_{\text{reg}}]$); the dosing interval of the first dose adjustment (m_0, see Remark 3.1 below).
The individual realizations of the random-effect components are unknown.

Initial step: The first dose D_0 is computed by the I/O formula (3.12). The output injection matrix $\mathcal{L}$ and the discrete-time state estimator are determined. For all $i \in \{0, \ldots, m_0 - 1\}$, $D_i = D_0$. At time $m_0 T$, we define

$$GAP := \alpha - C_{\text{reg}}\tilde{x}[m_0 - 1],$$

the "gap" at the beginning of the m_0th dosing interval.

Intermediate step: The model is updated after EBEs. The output injection matrix $\mathcal{L}$ and the discrete-time state estimator are updated, as well as the time between two adjustments of the dose (see Remark 3.1). Replacing $\alpha \leftarrow GAP$ in (3.12) yields the following estimated state feedback law:

$$D_1 = D_0 + \widetilde{D},$$
$$D_i = D_1$$

for all $i \in \{m_0, \ldots, m_1 - 1\}$, where

$$\widetilde{D} = \frac{GAP \cdot \Delta}{\sum_{i=1}^{n} \frac{F_i}{\lambda_i} \frac{1 - e^{-\lambda_i \Delta}}{e^{-\lambda_i T} - 1}}, \tag{3.15}$$

where, for all $i \in \{1, \ldots, n\}$, λ_i is the ith eigenvalue of the updated state matrix A and $F_i = \lim_{s \to \lambda_i} \left(\hat{g}(s)\,(s - \lambda_i)\right)$ denotes the residual in λ_i of the transfer function $\hat{g}$ corresponding to the system $[A, b, C_{\text{reg}}]$.
$q \leftarrow 1$.

Recurrence step: While $|GAP| \geq 0.01$, let

$$GAP := \alpha - C_{\text{reg}}\tilde{x}[m_q - 1].$$

The dose is updated as follows:

$$D_q = D_{q-1} + \widetilde{D},$$
$$D_i = D_q$$

for all $i \in \{m_{q-1}, \ldots, m_q - 1\}$, where $\widetilde{D}$ is given by (3.15). $q \leftarrow q + 1$.

The block diagram in Fig. 3.7 represents the resulting closed-loop system.

Remark 3.1. The dose adjustment (state feedback) should be performed at steady state. In this work, we assume that the steady state is reached after five time constants τ related to the dominant mode ($\tau := |1/\lambda_F|$, where λ_F is the Frobenius eigenvalue [10]). The dominant mode has then reached a value less than 1% of its initial value. The concept of time constant is

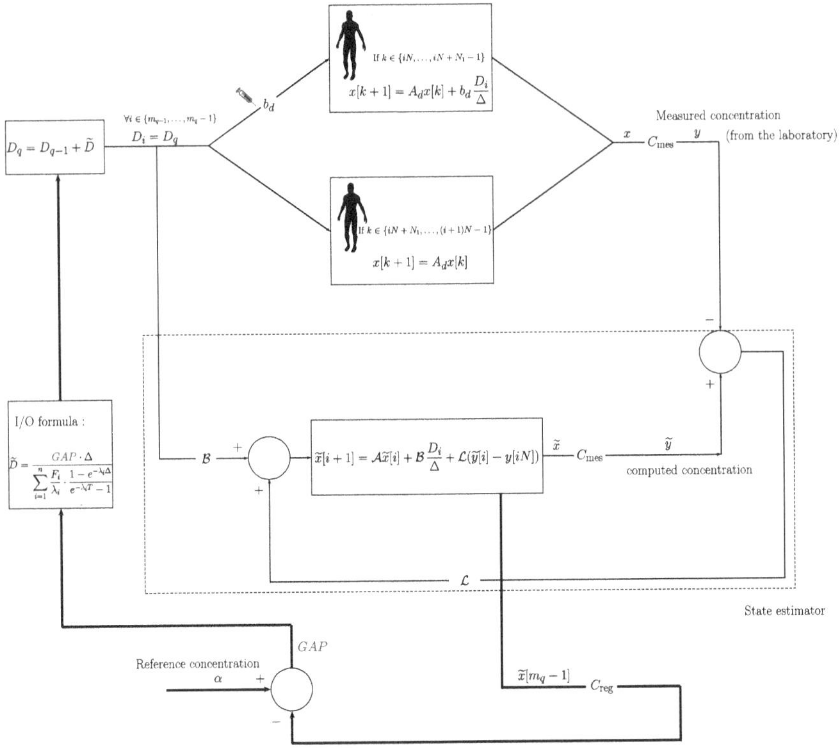

FIGURE 3.7 Block diagram representing the closed-loop system (recurrence step, $q \in \mathbb{N}_0$). $[A_d, b_d, C_{mes}]$, (discrete-time) model specific to the individual patient. The coefficients F_i and λ_i $(i \in \{1, 2, 3, \ldots n\})$ are directly related to the continuous-time model specific to the individual patient $[A, b, C_{reg}]$).

close to the PK concept of half-life (see, e.g., [26]). Eventually, the relevant dosing intervals to update the dose are estimated by

$$\begin{cases} m_0 = \lfloor 5\tau_0/T \rfloor + 1, \\ m_q = m_{q-1} + \lfloor 5\tau/T \rfloor, \quad \forall q \in \mathbb{N}_0, \end{cases}$$

where τ_0 is the time constant of the nominal model (i.e., before the EBEs) and τ is the time constant of the updated model (i.e., after the EBEs).

Let us consider a virtual patient with a body weight of 75 kg and a GFR of 65 mL/min. Figs. 3.8a and 3.8b show the input function and the PK profile attributable to this feedback control law. The initial AB treatment was ~1.1 g over 3 hours every 8 hours. The Bayesian analyses to recalculate the model parameters were performed at the beginning of the fourth dosing interval (m_0). The time constant was then reevaluated for the updated model. The successive doses were 5.18 g, 4.60 g, and 3.94 g, updated

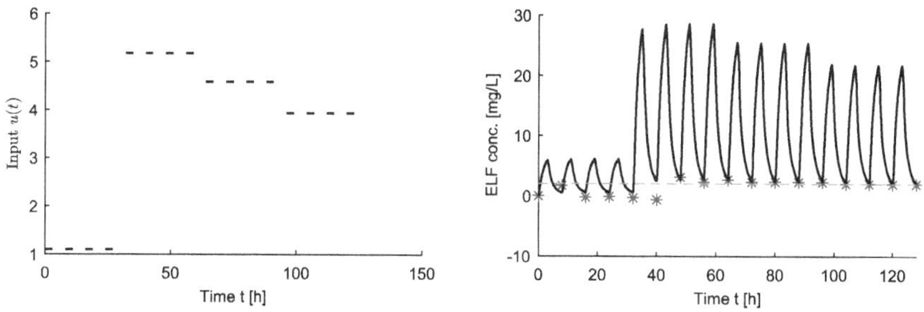

FIGURE 3.8 Illustration of the estimated state feedback algorithm. The blue (dark gray in print version) points correspond to the estimated state corrected by the scaling factor.

every four infusions. In this illustration, the control law was based on the estimated state at the infection site, i.e.,

$$C_{\text{reg}} = \begin{pmatrix} 0 & 1/S_2 & 0 & \cdots & 0 \end{pmatrix},$$

while the measures were plasma samples, as indicated by C_{mes}.

3.5 Conclusion

System and control theory provides tools to address issues in clinical pharmacology, including dosing for clinical care of (critically ill) patients. In this chapter, related to time-dependent ABs, we have described approaches to compute and update the dose/infusion rate of each administration, assuming that the infusion duration and the length of the dosing interval are fixed (previous clinician's decision).

Currently, the published model-based methods for AB dosing are mostly based on Monte Carlo simulations and related (systemic) PTA, such that they provide dosing recommendations at a population level. In the input-output approach presented here, it is no longer required to compare different dose levels by numerical simulations. We have derived a formula in order to express the dose with respect to the infusion duration Δ and the dosing interval T, allowing to reach a predefined concentration (in the plasma or at the infection site) for each patient of interest, given their relevant characteristics (covariates). This input-output formula can then be incorporated into a worst-case analysis (which computes a conservative value, but takes into account the variability in drug concentrations) or be used to design an estimated state feedback that is based on measurements of plasma concentration equally spaced in time. The drug dosing strategies derived here should be interpreted as decision-making aids providing guidelines to physicians. Meropenem was presented as a case study. Through numerical simulations, the methods were successfully applied with acceptable robustness and reliable results.

Except for the system analysis (Section 3.2.2) which is specific to compartmental popPK models, all the results can be applied, e.g., to PBPK models. Such models are good candidates as they are more robust to predict concentrations in different populations and settings: the model structure and parameters are less dependent on the available data because they are based on pathophysiology considerations and the drug's pharmacological properties (see [25] and references therein).

Acknowledgments

The content reported in this chapter has been communicated at the 1st Virtual IFAC World Congress (IFAC-V 2020), Berlin, Germany.

References

[1] J.M. Bailey, W.M. Haddad, Drug dosing control in clinical pharmacology, IEEE Control Syst. Mag. 25 (2) (April 2005) 35–51.

[2] R.J. Bauer, NONMEM tutorial part I: description of commands and options, with simple examples of population analysis, CPT Pharmacometrics Syst. Pharmacol. 8 (8) (2019) 525–537, https://doi.org/10.1002/psp4.12404.

[3] M. Beumier, G.S. Casu, et al., Elevated β-lactam concentrations associated with neurological deterioration in ICU septic patients, Minerva Anestesiol. (ISSN 0375-9393) (2015).

[4] V. Chellaboina, et al., Modeling and analysis of mass-action kinetics, IEEE Control Syst. Mag. 29 (4) (Aug. 2009) 60–78.

[5] M.O. Cotta, J.A. Roberts, J. Lipman, Antibiotic dose optimization in critically ill patients, Med. Intensiva (ISSN 1578-6749) 39 (9) (2015) 563–572, https://doi.org/10.1016/j.medin.2015.07.009.

[6] F. Frippiat, F.T. Musuamba, et al., Modelled target attainment after meropenem infusion in patients with severe nosocomial pneumonia: the promesse study, J. Antimicrob. Chemother. 70 (1) (2015) 207–216.

[7] J. Goncalves-Pereira, et al., Assessment of pharmacokinetic changes of meropenem during therapy in septic critically ill patients, BMC Pharmacol. Toxicol. 15 (1) (2014) 15–21.

[8] W.M. Haddad, V. Chellaboina, Q. Hui, Nonnegative and Compartmental Dynamical Systems, Princeton University Press, 2010.

[9] A. Haidar, The artificial pancreas: how closed-loop control is revolutionizing diabetes, IEEE Control Syst. Mag. 36 (5) (2016) 28–47.

[10] R. Horn, C. Johnson, Matrix Analysis, Cambridge University Press, 1985.

[11] S. Jaruratanasirikul, S. Thengyai, W. Wongpoowarak, et al., Population pharmacokinetics and Monte Carlo dosing simulations of meropenem during the early phase of severe sepsis and septic shock in critically ill patients in intensive care units, Antimicrob. Agents Chemother. 59 (6) (2015) 2995–3001, https://doi.org/10.1128/AAC.04166-14.

[12] S. Kiem, J.J. Schentag, Interpretation of antibiotic concentration ratios measured in epithelial lining fluid, Antimicrob. Agents Chemother. 52 (1) (2008) 24–36, https://doi.org/10.1128/AAC.00133-06.

[13] C. Li, J.L. Kuti, C.H. Nightingale, et al., Population pharmacokinetic analysis and dosing regimen optimization of meropenem in adult patients, J. Clin. Pharmacol. 46 (10) (2006) 1171–1178.

[14] L. Lindbom, P. Pihlgren, N. Jonsson, PsN-Toolkit - a collection of computer intensive statistical methods for non-linear mixed effect modeling using NONMEM, Comput. Methods Programs Biomed. 79 (3) (2005) 241–257, https://doi.org/10.1016/j.cmpb.2005.04.005.

[15] T.P. Lodise, et al., Penetration of meropenem into epithelial lining fluid of patients with ventilator-associated pneumonia, J. Antimicrob. Chemother. 55 (4) (2011) 1606–1610.

[16] F. Mattioli, C. Fucile, D. Del Bono, et al., Population pharmacokinetics and probability of target attainment of meropenem in critically ill patients, Eur. J. Clin. Pharmacol. 72 (7) (2016) 839–848.

[17] MHRA, Medicines and healthcare products regulatory agency. Meropenem trihydrate, in: Medicines Information: SPC & PILs, 2019, http://www.mhra.gov.uk/spc-pil/. (Accessed 20 March 2020).

[18] D.R. Mould, R.N. Upton, Basic concepts in population modeling, simulation, and model-based drug development, CPT Pharmacometrics Syst. Pharmacol. 1 (9) (2012) 1–14.

[19] D.R. Mould, R.N. Upton, Basic concepts in population modeling, simulation, and model-based drug development-part 2: introduction to pharmacokinetic modeling methods, CPT Pharmacometrics Syst. Pharmacol. 2 (4) (2013) 1–14.

[20] F.T. Musuamba, E. Manolis, N. Holford, et al., Advanced methods for dose and regimen finding during drug development: summary of the EMA/EFPIA workshop on dose finding (London 4-5 December 2014), CPT Pharmacometrics Syst. Pharmacol. 6 (7) (2017) 418–429, https://doi.org/10.1002/psp4.12196.

[21] T. Nguyen, et al., Model evaluation of continuous data pharmacometric models: metrics and graphics, CPT Pharmacometrics Syst. Pharmacol. 6 (2) (2017) 87–109.

[22] J.S. Owen, J. Fiedler-Kelly, Introduction to Population Pharmacokinetic/Pharmacodynamic Analysis with Nonlinear Mixed Effects Models, John Wiley & Sons, 2014.

[23] F.B. Sime, M.S. Roberts, J.A. Roberts, Optimization of dosing regimens and dosing in special populations, Clin. Microbiol. Infect. (ISSN 1469-0691) 21 (10) (2015) 886–893, https://doi.org/10.1016/j.cmi.2015.05.002.

[24] F. Sjövall, et al., Maximally effective dosing regimens of meropenem in patients with septic shock, J. Antimicrob. Chemother. (ISSN 1460-2091) 73 (1) (2018) 191–198, https://doi.org/10.1093/jac/dkx330.

[25] P. Thémans, P. Marquet, J.J. Winkin, F.T. Musuamba, Towards a generic tool for prediction of meropenem systemic and infection-site exposure: a physiologically based pharmacokinetic model for adult patients with pneumonia, Drugs R D 19 (2) (2019) 177–189, https://doi.org/10.1007/s40268-019-0268-x.

[26] P.L. Toutain, A. Bousquet-Mélou, Plasma terminal half-life, J. Vet. Pharmacol. Ther. 27 (6) (2004) 427–439.

[27] M. Usman, O.R. Frey, G. Hempel, Population pharmacokinetics of meropenem in elderly patients: dosing simulations based on renal function, Eur. J. Clin. Pharmacol. 73 (3) (2017) 333–342, https://doi.org/10.1007/s00228-016-2172-4.

[28] R.P. Veiga, J.-A. Paiva, Pharmacokinetics–pharmacodynamics issues relevant for the clinical use of beta-lactam antibiotics in critically ill patients, Crit. Care 22 (1) (2018) 233, https://doi.org/10.1186/s13054-018-2155-1.

[29] A.A. Vinks, The application of population pharmacokinetic modeling to individualized antibiotic therapy, Int. J. Antimicrob. Agents 19 (4) (2002) 313–322.

[30] S.G. Wicha, TDMx - model-supported Therapeutic Drug Monitoring for Precision Dosing, 2018, http://www.tdmx.eu/. (Accessed 28 October 2019).

CHAPTER

4

Safe glycemia regulation considering parameter variations under the offset-free MPC with pulse inputs scheme

María F. Villa-Tamayo and Pablo S. Rivadeneira

Universidad Nacional de Colombia, Facultad de Minas, Grupo GITA, Medellín, Colombia

4.1 Introduction

Control systems with pulse inputs describe systems in which a discontinuous control action has an amplitude with nonnegligible duration but remains null a significant part of the sampling time. Therefore, the system has two stages in a period: a forced response while the input is operating and a free response while the input is zero. A special case of pulse systems are those with impulsive inputs which have been mostly studied for applications such as drug scheduling [1,2]. When the administration of medications has a very short duration in relation to the sampling period, for instance, an injection or a pill in one day, it can be modeled as an impulse (Dirac delta). Different medical treatments have been addressed as impulsive systems including type 1 diabetes mellitus (T1DM) [3], oncolytic virus therapy [4], and human immunodeficiency virus (HIV) [5].

However, given the technological advances in medical devices, which allow a more intensive control of the disease by means of the input application every certain period, the treatment is better modeled as a pulse-controlled system. This is the case of the well-known artificial pancreas for T1DM treatment. An artificial pancreas consists of a continuous glucose monitor (CGM), a continuous insulin infusion pump (some pumps

Feedback Control for Personalized Medicine
https://doi.org/10.1016/B978-0-32-390171-0.00013-5

on the market allow insulin infusion every 5–15 minutes), and a control strategy to calculate the appropriate dose according to glucose measurements [6,7]. Then, in a predefined sampling period for the pump, insulin injection occurs with a certain duration but without covering the entire period, i.e., the input acts as a pulse.

For systems with pulse inputs, a first approach to a model predictive control (MPC) strategy was introduced by [8]. Nevertheless, mathematical characterizations and different control strategies have been more substantiated for the specific case of impulsive systems. Among these, a feedback control was proposed in [9], MPC strategies were developed by [1,10,11], and an offset-free MPC formulation was presented by [12,13], which was based on the offset-free MPC strategy for discrete-time systems [14–16] to compensate the effect of a plant–model mismatch.

MPCs have reached large popularity for biomedical systems, likely because they can handle system constraints more easily [6,11,17–19]. MPC strategies solve a finite-horizon optimization problem at every sampling time to obtain an optimal profile of inputs that minimizes an objective function, constrained to a model describing the process evolution. The cost function is designed to achieve control goals, while in addition, other constraints on the control action and state, usually given by physical limitations of the process or to reflect a required performance, are usually imposed onto the minimization problem. From the input profile, the first element is applied to the plant, and when a new process measurement is available, the optimization problem is repeated. More details about MPC formulations can be seen in [20].

In this chapter, the offset-free MPC formulation for systems with pulse inputs is presented as a generalization to the works developed for impulsive and discrete systems. The main characteristics of the strategy here proposed are: (i) the use of a zone MPC (ZMPC) strategy with artificial equilibrium variables which provides an enlarged domain of attraction, (ii) an augmented system with a disturbance model to compensate for the effect of parametric variations of the plant that lead to a plant–model mismatch, (iii) the incorporation of the estimation of the mismatch in the MPC constraints, and (iv) the calculation of the pulse control action including its equilibrium characterization.

The offset-free strategy is evaluated in T1DM treatment where it is intended to maintain the blood glucose (BG) concentration in a target range. Then, to illustrate another advantage of the offset-free feature, the control strategy is enhanced with a safety layer (SL) for the insulin-on-board (IOB) by using the estimation of the plant–model mismatch. Control formulations based on IOB are guided to prevent hypoglycemic events caused by insulin stacking. The authors in [21,22] developed a hypoglycemia safety condition based on IOB constraints that can be implemented separately from the main controller. For MPC formulations, the IOB constraint was first introduced as a dynamic boundary with estimated insulin action decay curves in [23]. This idea was also incorporated in the strategy presented in [7], where the IOB boundary changes as a function of glycemia. Another approach is the one proposed in [24] where an IOB decaying curve is used to set the input reference in the MPC cost function. However, although these developments for safety control are effective to avoid hypoglycemia, constraining the IOB can be harmful when there are physiological variations in the subject that induce hyperglycemia.

Therefore, to counteract this last problem, the layer based on IOB estimation is combined with the offset-free MPC controller by setting specific rules to adapt the SL. It will be shown how the information on the plant–model mismatch describes the direction of physiological variations: towards hypo- or hyperglycemia. Therefore, the rules are established based on the sign of the estimated plant–model mismatch to tighten or release the IOB constraint when detecting variations that induce hypo- or hyperglycemia, respectively.

The control strategies for T1DM treatment are tested in a cohort of 10 virtual adult subjects from the UVA/Padova simulator [25]. To show the benefits of the strategy, simulations are performed when there is a plant–model mismatch caused by variations in the parameters and also sensor noise is considered. In real life, the variations occur in the patient due to physical activity, stress, and hormonal changes, among others [17]. Satisfactory results are obtained by compensating the effect of parametric variations in BG levels, by steering glycemia to the target and reducing hypoglycemia events.

4.2 System with pulse inputs

Consider the following state-space representation of a continuous affine linear system:

$$\begin{aligned} \dot{x}(t) &= Ax(t) + B_u u(t) + B_r r(t) + E, \\ y(t) &= Cx(t), \end{aligned} \tag{4.1}$$

where $\mathrm{x} \in \mathrm{X} \subset \mathbb{R}^{\mathrm{n_x}}$, $\mathrm{u} \in \mathrm{U} \subset \mathbb{R}^{n_u}$, and $r \in R \subseteq R^{n_r}$ are the constrained state, constraint control inputs, and disturbances, respectively. Constraint sets contain the origin in their interior, X is closed, and U, R are compact. Matrix E is a constant term.

As the interest consists of systems with pulse inputs, let the control input u be the signal of form:

$$u(t) = \begin{cases} u(kT), & t \in [kT, kT + \Delta T], \\ 0, & t \in [kT + \Delta T, (k+1)T], \end{cases} \tag{4.2}$$

where T is a fixed time period and ΔT represents the duration of the input signal inside period T (the pulse width in T). Then, by assuming a constant disturbance in each sampling period ($r(t) = r(KT)$), the solution $\phi(t, x(kT), u(\cdot), r(\cdot))$ of (4.1) for $t \in [kT, (k+1)T]$ is

$$\phi(t, x(kT), u(\cdot), r(\cdot)) = e^{A(t-kT)} x(kT) + \int_{kT}^{t} e^{A(t-\zeta)} B_u u(\zeta) d\zeta + \int_{kT}^{t} e^{A(t-\zeta)} d\zeta B_r r(kT) + \int_{kT}^{t} e^{A(t-\zeta)} d\zeta E. \tag{4.3}$$

Thus, as observed in the illustrative example of Fig. 4.1, the system is characterized by two responses: a forced response when applying the input pulse of duration ΔT and a free response when the input is zero during the remaining time of the period $T - \Delta T$. Next, the idea is to characterize the state by sampling it at times kT ($x((k+1)T) = \phi((k+1)T, x(kT), u(\cdot), r(\cdot))$) and $kT + \Delta T$ ($x((k+1)T + \Delta T) = \phi((k+1)T + \Delta T, x(kT + \Delta T), u(\cdot), r(\cdot))$) [8], with the aim of applying control strategies that have been widely developed for discrete-time systems.

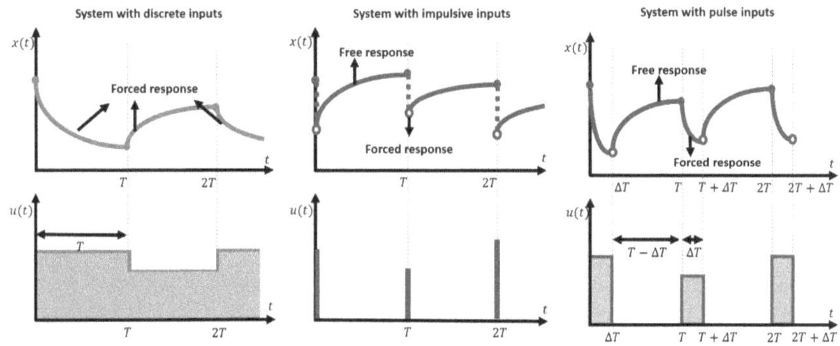

FIGURE 4.1 Evolution of the system when applying different types of inputs. Left: discrete inputs are applied, and thus, there is no free response. Middle: impulsive inputs are applied, and thus, there are discontinuities or jumps at the impulse times and a free response during period T. Right: pulse inputs of duration ΔT are applied generating a forced response, and the system evolves autonomously during the remaining time $T - \Delta T$.

4.2.1 Sampled state at times kT

Since matrices A, B_u, B_r, and E are assumed time-invariant, the integrals in (4.3) can be considered in time $[0 \quad T]$. Then, the sampled state is given by

$$x((k+1)T) = e^{AT} x(kT) + \int_{0}^{T} e^{A(T-\zeta)} B_u u(\zeta - kT) d\zeta \tag{4.4}$$

$$+\int_0^T e^{A(T-\zeta)} d\zeta\, B_r r(kT) + \int_0^T e^{A(T-\zeta)} d\zeta\, E.$$

In addition, the input effect over the state can be rewritten by following the form of u in (4.2):

$$\begin{aligned}\int_0^T e^{A(T-\zeta)} B_u u(\zeta - kT)\, d\zeta &= \int_0^{\Delta T} e^{A(T-\zeta)} d\zeta\, B_u u(kT) \\ &= e^{A(T-\Delta T)} \int_0^{\Delta T} e^{A(\Delta T-\zeta)} d\zeta\, B_u u(kT) \\ &= e^{A(T-\Delta T)} B_u^{\Delta T} u(kT).\end{aligned} \tag{4.5}$$

Then, by denoting each integral term in (4.4) as $\mathrm{A^d} = \mathrm{e^{AT}}$, $B_u^{d_1} = e^{A(T-\Delta T)} B_u^{\Delta T}$, $B_r^d = \int_0^T e^{A(T-\zeta)} d\zeta\, B_r$, and $E^d = \int_0^T e^{A(T-\zeta)} d\zeta\, E$ and the state, input, and disturbance at times kT as $x^\bullet(k)$, $u^\bullet(k)$, and $r^\bullet(k)$, respectively, the sampled expression is

$$x^\bullet(k+1) = A^d x^\bullet(k) + B_u^{d_1} u^\bullet(k) + B_r^d r^\bullet(k) + E^d. \tag{4.6}$$

4.2.2 Sampled state at times $kT + \Delta T$

By following a similar procedure, the state is sampled at the times when the free response starts:

$$\begin{aligned}x((k+1)T + \Delta T) = e^{AT} x(kT + \Delta T) + \int_0^{\Delta T} e^{A(T-\zeta)} d\zeta\, B_u u((k+1)T) \\ + \int_0^T e^{A(T-\zeta)} d\zeta\, B_r r(kT) + \int_0^T e^{A(T-\zeta)} d\zeta\, E,\end{aligned} \tag{4.7}$$

which can be rewritten as

$$x^\circ(k+1) = A^d x^\circ(k) + B_u^{d_2} u^\circ(k) + B_r^d r^\circ(k) + E^d, \tag{4.8}$$

where $A_d = e^{AT}$, $B_u^{d_2} = \int_0^{\Delta T} e^{A(T-\zeta)} d\zeta\, B_u = B_u^{\Delta T}$, $B_r^d = \int_0^T e^{A(T-\zeta)} d\zeta\, B_r$, and $E^d = \int_0^T e^{A(T-\zeta)} d\zeta\, E$ and the state, input, and disturbance at times $kT + \Delta T$ are denoted by $x^\circ(k)$, $u^\circ(k)$, and $r^\circ(k)$, respectively.

Lastly, note that the system does not have an equilibrium point because of the combination of both forced and free responses that result from applying the pulse input. Hence, an extended equilibrium set is described given by the equilibrium of systems (4.6) and (4.8) and by including the continuous-time evolution between each point:

$$o_s\left(x_s^\bullet, u_s^\bullet\right) = \left\{\phi\left(t, x_s^\bullet, u_s^\bullet(t)\right), t \in [kT, (k+1)T], k \in \mathbb{N}\right\},$$

which must be feasible. Therefore, the equilibrium sets are

$$X_s^\bullet = \{x_s^\bullet \in X : \exists u_s^\bullet \in U \text{ such that } x_s^\bullet = A^d x_s^\bullet + B_u^{d_1} u_s^\bullet + E^d, o_s\left(x_s^\bullet, u_s^\bullet\right) \in X\},$$
$$X_s^\circ = \{x_s^\circ \in X : \exists u_s^\circ \in U \text{ such that } x_s^\circ = A^d x_s^\circ + B_u^{d_2} u_s^\circ + E^d\}. \quad (4.9)$$

Although both subsystems (4.6) and (4.8) are required to characterize the equilibrium region, note that the pulse input is applied to $x^\bullet$, which is also the state directly related to the measurable variable. Thus, the sampled model (4.6) is enough to be used in the control strategy.

4.3 Offset-free MPC strategy

When the prediction model utilized in the control strategy is not accurate enough to describe the plant behavior, i.e., when there is a plant–model mismatch, the MPC can generate a misleading control action that does not steer the system to the target and normally generates a steady-state offset.

In this section, the aim is to develop a control strategy that drives the system to a nonempty target set by means of pulse control actions while satisfying constraints and compensating for the effect of a constant plant–model mismatch. To that purpose, the offset-free ZMPC (denoted as ZMPC-OF) previously developed for discrete systems [14–16] and for impulsive systems [13] is here generalized to the pulse scheme. The idea of this strategy is to obtain some information about the mismatch and provide it to the ZMPC so that it corrects its prediction model and target.

To that end, the nominal model of the plant is augmented with a model that considers the disturbances in the form of an integrating state ($\dot{d}(t) = 0$). Denote $\tilde{x}^\bullet = [x^\bullet d^\bullet]'$ as the augmented state, where the disturbance $d \in \mathbb{R}^{n_d}$ has associated matrices $B_d \in \mathbb{R}^{n_x \times n_d}$ and $C_d \in \mathbb{R}^{n_y \times n_d}$. Then, the extended system and output of the model with pulse input has the form

$$\tilde{x}^\bullet(k+1) = \tilde{A}^d \tilde{x}^\bullet(k) + \tilde{B}_u^{d_1} u^\bullet(k) + \tilde{B}_r^d r^\bullet(k) + \tilde{E}^d, \quad (4.10)$$
$$\tilde{y}(k) = \tilde{C}\tilde{x}^\bullet(k),$$

where $\tilde{A}^d = \begin{bmatrix} A^d & B_d^d \\ 0 & I \end{bmatrix}$, $\tilde{B}_u^{d_1} = \begin{bmatrix} B_u^{d_1} \\ 0 \end{bmatrix}$, $\tilde{B}_r^d = \begin{bmatrix} B_r^d \\ 0 \end{bmatrix}$, $\tilde{E}^d = \begin{bmatrix} E^d \\ 0 \end{bmatrix}$, $\tilde{C} = [C \quad C_d]$, and $B_d^d = e^{AT} B_d$.

To consider the mismatch in the optimization problem, the state and disturbance have to be estimated. Therefore, it is necessary to select matrices B_d and C_d to ensure that the augmented model is observable. This is obtained if and only if

$$rank \begin{bmatrix} \tilde{C} \\ \tilde{C}\tilde{A}^d \\ \vdots \\ \tilde{C}\tilde{A}^{d^{n_x+n_d-1}} \end{bmatrix} = n_x + n_d, \tag{4.11}$$

and then, an estimator for the state and disturbance is designed in the form

$$\hat{\tilde{x}}^{\bullet}(k+1) = \tilde{A}^d \hat{\tilde{x}}^{\bullet}(k) + \tilde{B}_u^{d_1} u^{\bullet}(k) + \tilde{B}_r^d r^{\bullet}(k) + \tilde{E}^d + L\left(y(k) - \tilde{C}\hat{\tilde{x}}^{\bullet}(k)\right), \tag{4.12}$$

where matrix L is chosen so that the estimator is stable. Here, the Kalman filter algorithm is used to obtain the estimation.

The offset-free control strategy consists of a baseline MPC and the offset-free component. The baseline formulation selected is the ZMPC developed for impulsive systems in [1] and extended to pulse systems in [8]. Its main idea is to steer the system from an initial point to a target set X^{Tar} through the equilibrium points using artificial/intermediary steady-state variables $(x_a, u_a) \in X_s^{\bullet}, U_s^{\bullet}$. The cost function of the problem is

$$V_N = V_{dyn}\left(x^{\bullet}; \boldsymbol{u}, x_a, u_a\right) + V_f\left(X_s^{\bullet Tar}, U_s^{Tar}; x_a, u_a\right), \tag{4.13}$$

which is composed of two sections: (i) the dynamic cost,

$$V_{dyn}\left(x^{\bullet}; \boldsymbol{u}, x_a, u_a\right) = \sum_{j=0}^{N-1} \| x^{\bullet}(j) - x_a \|_Q^2 + \sum_{j=0}^{N-1} \| u^{\bullet}(j) - u_a \|_R^2,$$

which steers the state to the artificial equilibrium inside $\left(X_s^{\bullet}, U_s^{\bullet}\right)$, and (ii) the terminal cost,

$$V_f\left(X_s^{\bullet Tar}, U_s^{Tar}; x_a, u_a\right) = P\left(dist_{X_s^{\bullet Tar}}(x_a) + dist_{U_s^{Tar}}(u_a)\right),$$

which forces the artificial variables to an equilibrium that maintains the output $y(t)$ in the target set $Y^{Tar} = CX_s^{\bullet Tar}$, where $dist_Z(x)$ denotes the distance from a point x to set Z and $X_s^{\bullet Tar}$ is the maximum equilibrium set of the sampled model (4.4), such that its equilibrium trajectories remain inside X^{Tar}.

Next, this baseline formulation is merged with the offset-free feature by using the augmented model as the prediction model. Given the current estimate of the augmented state $\hat{\tilde{x}}^{\bullet}$, the optimization problem that solves the ZMPC-OF with pulse input every time k is

$$\min_{\boldsymbol{u}, x_a, u_a} \quad V_N\left(x^{\bullet}, X_s^{\cdot Tar}, U_s^{Tar}; \boldsymbol{u}, x_a, u_a\right)$$

$$
\begin{aligned}
s.t.\quad & x^{\bullet}(0) = \hat{x}^{\bullet}(k), \qquad d^{\bullet}(0) = \hat{d}^{\bullet}(k), \\
& x^{\bullet}(j+1) = A^{d} x^{\bullet}(j) + B_u^{d_1} u^{\bullet}(j) + B_r^{d} r^{\bullet}(j) + B_d^{d} d^{\bullet}(j) + E^{d}, \\
& d^{\bullet}(j+1) = d^{\bullet}(j), \\
& u^{\bullet}(j) \in U, \qquad x^{\bullet}(j) \in X, \\
& x^{\bullet}(N) = x_a, \\
& y_a = C x_a + C_d d^{\bullet}(j), \\
& x_a = A^{d} x_a + B_u^{d_1} u_a + B_d^{d} d^{\bullet} + E^{d},
\end{aligned}
\tag{4.14}
$$

where the first element of the solution sequence $\boldsymbol{u} = \{u^{\bullet}(0;x), u^{\bullet}(1;x), \ldots, u^{\bullet}(N-1;x)\}$ is applied to the plant every time k. The constraint $x^{\bullet}(N) = x_a$ in (4.14) forces the state at the end of the horizon to reach the artificial equilibrium x_a. Note that the plant–model mismatch is taken into account in both the prediction model and in the constraints $x_a = A^d x_a + B_u^{d_1} u_a + B_d^d d^{\bullet} + E^d$ and $y_a = C x_a + C_d d^{\bullet}$ to compute the artificial variables.

4.4 Application to T1DM treatment

In this section, the control strategy is applied in the context of artificial pancreas. The objective of the T1DM treatment is to maintain the BG concentration in a healthy zone known as normoglycemia (BG $\in$ [70–140] mg/dl during the fasting period, BG $\in$ [70–180] mg/dl during the postprandial periods) and avoid hyperglycemia (BG >180 mg/dl) and hypoglycemia (BG < 70 mg/dl) events. The selection of the target set is also part of the control design process; thus, it has been chosen inside the normoglycemia zone as $Y^{Tar} = \{y : 90 \le y \le 110\}$ to prevent the system from easily leaving the healthy zone when there are disturbances. A sampling time of $T = 5$ min is used, and the duration for insulin boluses (delivered by the insulin pump) is set as $\Delta T = 1$ min.

The model used to represent the glycemia–insulin–carbohydrate dynamics is a modified version of the minimal model developed in [26] to guarantee the structural identifiability [27]. The model has a continuous affine linear form (4.1) with corresponding matrices given by

$$
A = \begin{bmatrix} \theta_0 & -\theta_1 & 0 & \theta_2 & 0 \\ 0 & -\frac{1}{\theta_4} & \frac{1}{\theta_4} & 0 & 0 \\ 0 & 0 & -\frac{1}{\theta_4} & 0 & 0 \\ 0 & 0 & 0 & -\frac{1}{\theta_5} & \frac{1}{\theta_5} \\ 0 & 0 & 0 & 0 & -\frac{1}{\theta_5} \end{bmatrix}, \quad B_u = \begin{bmatrix} 0 \\ 0 \\ \frac{1}{\theta_4} \\ 0 \\ 0 \end{bmatrix}, \quad B_r = \begin{bmatrix} 0 \\ 0 \\ 0 \\ 0 \\ \frac{1}{\theta_5} \end{bmatrix}, \quad E = \begin{bmatrix} \theta_3 \\ 0 \\ 0 \\ 0 \\ 0 \end{bmatrix},
\tag{4.15}
$$

and $C = [1 \quad 0 \quad 0 \quad 0 \quad 0]$. The five state variables are x_1, the glycemia (mg/dl); x_2 and x_3, the delivery rates of insulin in the blood and subcutaneous space compartments, respectively (U/min); and x_4 and x_5, the delivery rates of carbohydrates in the stomach and duodenum compartments, respectively (g/min). The inputs are u, the exogenous insulin (U/min), and r, the carbohydrate intake due to meals (g/min). The description of each parameter can be seen in Table 4.1. The state and input constraint sets are set as $X = \{x : [0 \quad 0 \quad 0] \leq x \leq [500 \quad 10 \quad 10]\}$ and $U = \{u : 0 \leq u \leq 7.5\}$, respectively. The set X is only defined for the controllable part of the system since state variables x_4, x_5 only depend on disturbance r.

TABLE 4.1 Model parameters.

Parameter	Description	Unit
θ_0	Hepatic autoregulation	1/min
θ_1	Insulin sensitivity	mg/dl/U
θ_2	Carbohydrate bioavailability	mg/dl/g
θ_3	Endogenous glucose production at zero-insulin level	mg/dl/min
θ_4	Time-to-maximum of effective insulin concentration	min
θ_5	Time-to-maximum appearance rate of glucose	min

Day-to-day variability within the same subject (intrapatient variability) continues to be one of the greatest challenges for the development of artificial pancreata. Day-to-day variations may be due to physical activity, stress, and hormonal changes, among others [17]; and they lead to an alteration in insulin requirements, which if not compensated can cause hyperglycemia or hypoglycemia events. To simulate physiological variations in the patient, the three parameters identified in sensitivity tests as the most influential in BG dynamics [12] are changed: θ_0, the hepatic autoregulation; θ_1, insulin sensitivity; and θ_3, endogenous glucose production at zero-insulin level. These parameters are varied so that (i) hypoglycemia is induced in the subject: θ_0 and θ_1 are increased 30% of their nominal values (the parameter values identified for each subject to best fit data from the UVA/Padova simulator) and θ_3 is reduced the same percentage, or (ii) hyperglycemia is induced: θ_0 and θ_1 are reduced 30% with respect to the nominal value and θ_3 is increased by the same percentage.

The first simulation scenario here presented consist of 15 hours, in which only one meal is provided to one subject at 5:00 with a carbohydrate content of 60 g to visualize postprandial and fasting evolution. In addition, variations towards hypoglycemia are considered. The results of applying the ZMPC and ZMPC-OF strategies in this scenario are depicted in Fig. 4.2. Blue (dark gray in print version) lines represent the nominal behavior when there is no plant–model mismatch and the ZMPC is used. It is observed that, after the postprandial time, BG concentration achieves the target set by applying the required basal. When there is a plant–model mismatch and the same control strategy is applied (red (gray in print version)

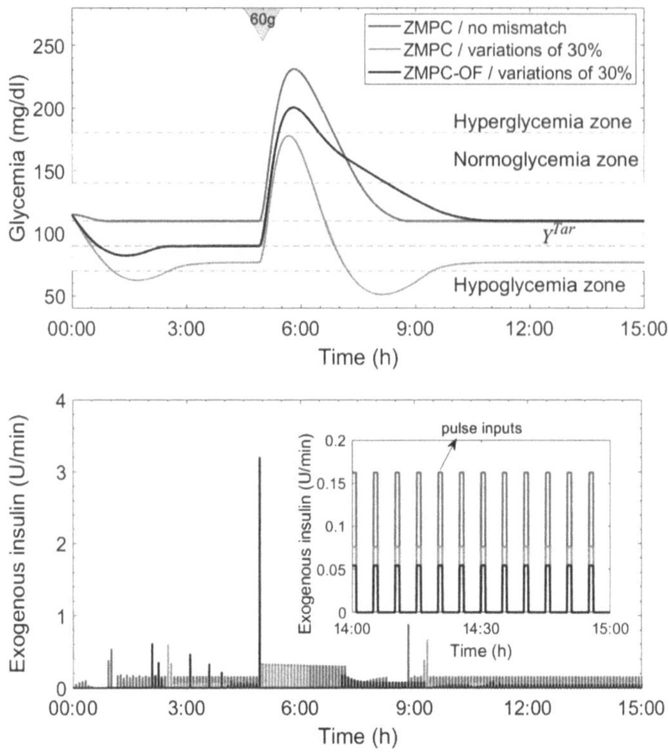

FIGURE 4.2 Glycemia evolution and exogenous insulin when applying control strategies with pulse inputs for adult patient number 6.

line), the ZMPC fails to adequately regulate glycemia, there are episodes of hypoglycemia, and at steady state, glycemia does not achieve the target zone, i.e., there is an offset. In contrast, when there is a mismatch and the ZMPC-OF is applied (black lines), insulin doses are corrected since the basal level is reduced compared to the one delivered by the ZMPC, and the glycemia is steered to the target Y^{Tar} while avoiding hypoglycemia events.

A second simulation scenario can be seen in Fig. 4.3, which shows the population results of 10 virtual adults identified from the UVA/Padova simulator. The simulation scenario consists of 3 days, and each day four meals are provided to each patient: 55 g at 7:00, 20 g at 10:00, 90 g at 13:00, and 70 g at 19:00. On days 1 and 3 metabolic variations were induced towards hypoglycemia and hyperglycemia, respectively, and day 2 was considered as the nominal case without mismatch to help eliminating any transition influence from the metabolic challenge of day 1 to day 3. The solid blue (dark gray in print version) line represents the population median when applying the ZMPC and the dotted black line represents the

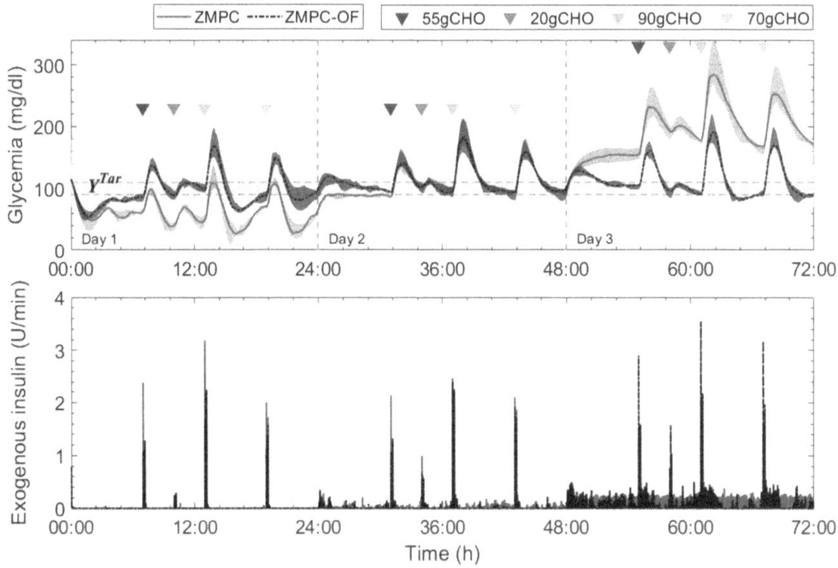

FIGURE 4.3 Glycemia evolution and exogenous insulin in 10 virtual adult patients when applying control strategies with pulse inputs and there is a plant–model mismatch. Solid (ZMPC) and dotted (ZMPC-OF) lines represent the median, shaded areas represent the interquartile range.

median with the ZMPC-OF. Shaded areas are the interquartile range of the population. One consideration is that the tuning of matrices Q and R has been established as aggressive, and it is the same tuning for both control strategies.

Given the plant–model mismatch with variations of 30% in parameters, the offset with the ZMPC is evident, at steady state the glycemia is steered to a point outside the target zone, and in the transitory at postprandial times, there are significant episodes of hypoglycemia and hyperglycemia on days 1 and 3, respectively. On the contrary, with the ZMPC-OF, the glycemia achieves the target zone, and hypoglycemia and hyperglycemia events are significantly reduced. From the insulin delivered to the system, it is shown how on day 1, insulin doses are reduced with the ZMPC-OF, and on day 2, the doses are increased with respect to those administered by the ZMPC. These changes are achieved thanks to the correction of the prediction model with the mismatch, allowing to supply the insulin doses required for each day according to the situation detected.

Additionally, the performance comparison between the two control strategies (ZMPC vs. ZMPC-OF) can be seen in detail in Table 4.2. Metrics shown are mean BG (mg/dl), standard deviation (SD) of BG (mg/dl), coefficient of variation (CV) of BG (%), time percentage of BG in each range (%), and number of events in range. Metrics are shown for each day of the

TABLE 4.2 Performance comparison between the system controlled with the ZMPC and with the ZMPC-OF in the 10-patient cohort.

	Day 1		Day 2		Day 3	
Strategy	**ZMPC**	**ZMPC-OF**	**ZMPC**	**ZMPC-OF**	**ZMPC**	**ZMPC-PF**
Mean BG (mg/dl)	59.9 ± 6.6	96.1(5.4)	108.8 ± 4.3	112.9 (4.2)	191.7 ± 15.3	114.3 ± 4.4
SD BG (mg/dl)	23.7 ± 4.4	27.2±5.8	24.4 (12.1)	23.6 ± 6.7	42.4 ± 9.7	28.2 ± 6.6
CV BG (mg/dl)	40.3 ± 10.4	29.0 ± 5.5	23.4 (9.4)	21.1 ± 5.5	22.0 ± 4.0	24.6 ± 5.1
Time percentage of BG in each range (%)						
<54 mg/dl	41.3 ± 10.5	4.5 ± 4.8	0 (0)	0 (0)	0 (0)	0 (0)
<70 mg/dl	71.8 ± 9.4	16.9 (4.7)	0.7 ± 0.7	0 (0)	0 (0)	0 (0)
70–140 mg/dl	27.9 ± 9.3	72.8 (10.2)	85.5 ± 4.7	85.3 (5.4)	6.5 (11.6)	82.5 ± 3.9
70–180 mg/dl	28.2 ± 9.4	81.4 (7.5)	96.6 ± 2.8	96.5 (6.0)	45.9 ± 12.4	96.9 (8.7)
>180 mg/dl	0 (0)	0 (2.8)	1.2 (6.2)	1.6 (4.8)	54.1 ± 12.4	3.9 ± 4.1
>250 mg/dl	0 (0)	0 (0)	0 (0)	0 (0)	11.5 ± 8.5	0 (0)
>300 mg/dl	0 (0)	0 (0)	0 (0)	0 (0)	0 (5.5)	0 (0)
Number of events in each range (-)						
<54 mg/dl	5.4 ± 1.3	1 (1)	0 (0)	0 (0)	0 (0)	0 (0)
<70 mg/dl	6.8 ± 1.0	2 (1)	0 (0)	0 (0)	0 (0)	0 (0)
>180 mg/dl	0 (0)	0 (1)	1 (2)	0.5 (1)	4 (0)	1 (2)
>250 mg/dl	0 (0)	0 (0)	0 (0)	0 (0)	2 (2)	0 (0)
>300 mg/dl	0 (0)	0 (0)	0 (0)	0 (0)	0 (1)	0 (0)

simulation scenario. The outcome indexes are reported as mean ±SD for normally distributed data and as median (interquartile range) otherwise [28]. For the critical days with parameter variations in each subject (days 1 and 2), the offset-free approach accomplishes improved control performance in terms of time percentage in normoglycemia [70–180 mg/dl] (day 1: 28.2 ± 9.4 vs. 81.4 [7.5], day 3: 45.9 ± 12.4 vs. 96.9 [8.7]), on day 1 there is a reduction of the time percentage in hypoglycemia (71.8 ± 9.4 vs. 16.9 [4.7]), and on day 3, the time in hyperglycemia is also reduced (54.1±12.4 vs. 3.9 ± 4.1).

4.5 Safety layer for insulin-on-board constraint

This section has two objectives. The first objective is to improve safety in the treatment of diabetes by including information regarding IOB, and the second is to show how the plant–model mismatch information can be used to incorporate new features into the controller or to monitor the system.

Insulin stacking is one of the major challenges in intensive insulin therapy. It is produced by an overreaction in the postprandial period, where when detecting high BG levels, higher insulin doses than those required are administered, thus leading to hypoglycemia between 3 to 5 hours after a meal [6]. In [22,29], an external layer to the control strategy that uses information of the IOB to reduce hypoglycemia events was developed. The basis of the strategy consists of modulating the control input generated by the main controller as a function of the IOB, so that the insulin that is still active in the body does not exceed a safety region delimited by an established boundary ($\overline{IOB}$). The SL consists of the following components: the estimation of the IOB, the boundary selection, and the computation of a gain γ to satisfy the IOB constraint.

4.5.1 Insulin-on-board estimation

A physiological definition for IOB is the amount of active insulin in the body from previous infusions. As explained in [30], this definition is represented as

$$IOB\,(t) = \int_0^t \left(u\,(\tau) - x_2\,(\tau)\right) d\tau.$$

Next, by using the model (4.15) a state function for the IOB is

$$IOB\,(t) = \theta_4\,(x_2\,(t) + x_3\,(t))\,, \tag{4.16}$$

where parameter θ_4 is related to the duration of insulin action. This last expression leads to an equivalent IOB definition, which is the amount of insulin in the plasma and the subcutaneous compartments from previous insulin boluses [29].

4.5.2 Insulin-on-board boundary

The IOB boundary considered here consists of β times the IOB profile of the open-loop treatment that would be applied to the subject [21,29]. This open-loop IOB profile ($\overline{IOB}$) is the time solution of the subsystem (x_2, x_3) in (4.16), when applying the basal u_b and the open-loop bolus $U_{CHO} = CHO/CR$, with CR and CHO denoting the carbohydrate-to-insulin ratio and the estimated carbohydrate content of the ingested meal, respectively. The constant CR can be provided either by the physician or derived from the model as $CR = p_2/p_1$. Hence, the boundary $\overline{IOB}$ is expressed as

$$\overline{IOB}\,(t) = \beta\left(U_{CHO}e^{-\frac{t}{\theta_4}}\left(\frac{t}{\theta_4}+1\right) + 2\theta_4 u_b\right). \tag{4.17}$$

By using this profile as constraint, it is ensured that the whole controller will not deliver much more insulin than the open-loop therapy. This is achieved by changing parameter β. When $\beta < 1$ the constraint is tightened, and thus, conservative control actions are obtained, and when $\beta > 1$ the constraint is released, allowing for more aggressive actions.

4.5.3 The switching signal

The idea of the SL is to predict the IOB evolution during the sampling period by running the algorithm with a shorter sampling time than the one of the insulin pump. Next, based on this prediction, a gain $\gamma \in [0, 1]$ is computed to modulate the signal u generated by the main controller (the ZMPC-OF in this case) to maintain the IOB inside the safety zone delimited by $\overline{IOB}$. Therefore, the final control action to be administered to the subject is

$$u_f(k) = \gamma(k)\, u(k).$$

The gain compatible with the IOB boundary is computed after running the following switching adaptation law [22]:

$$\omega(t) = \begin{cases} 1 & \text{if } \sigma(t) \geq 0, \\ 0 & \text{otherwise}, \end{cases} \tag{4.18}$$

where $\sigma(t) = \overline{IOB} - IOB(t)$ is constructed with the $IOB(t)$ estimator (4.16) and the IOB boundary determined in (4.17) and where $t \in [kT, (k+1)T)$. When $IOB(t)$ surpasses the boundary $\overline{IOB}$, then $\sigma < 0$, and thus, ω changes to 0 to cancel the insulin input until crossing the surface $\sigma = 0$. When $IOB(t)$ is below $\overline{IOB}$, $\sigma \geq 0$ and ω changes back to 1, allowing insulin delivery. Therefore, ω is switching at a very high frequency between 0 and 1. This effect is smoothed by computing the gain $\gamma(k)$ as the average

$$\gamma(k) = \frac{1}{n} \sum^{n} \omega(t),$$

where n is the number of samples of ω obtained during the pump updating period.

4.5.4 Coupling the safety layer with the offset-free strategy

Although the main advantage of the offset-free control is to correct the prediction model and target to eliminate the steady-sate offset produced when there is a plant–model mismatch, more advantages can be exploited by using the information provided throughout estimating the mismatch d. For the case of T1DM treatment, with this estimation it is possible to know when the subject's metabolic variations induce a state of either hypoglycemia or hyperglycemia [31].

Consider the 3-day simulation scenario in which variations in the subject push BG dynamics towards hypoglycemia on day 1 and towards hyperglycemia on day 3. As shown in Fig. 4.4, when applying the ZMPC-OF (black lines) there are some hypoglycemia events on day 1, but the offset has been corrected thanks to the estimation of d (also visualized in the figure). From this graph, it is observed that the sign of d is negative or positive depending on whether the net effect of variations is detected. On day 1, d showed a negative sign, indicating that hypoglycemia variations occurred, on day 2, d tended to zero, and on day 3, d showed a positive sign since the variations induce hyperglycemia. Thus, the idea is to use this information to achieve the best possible performance with the strategy.

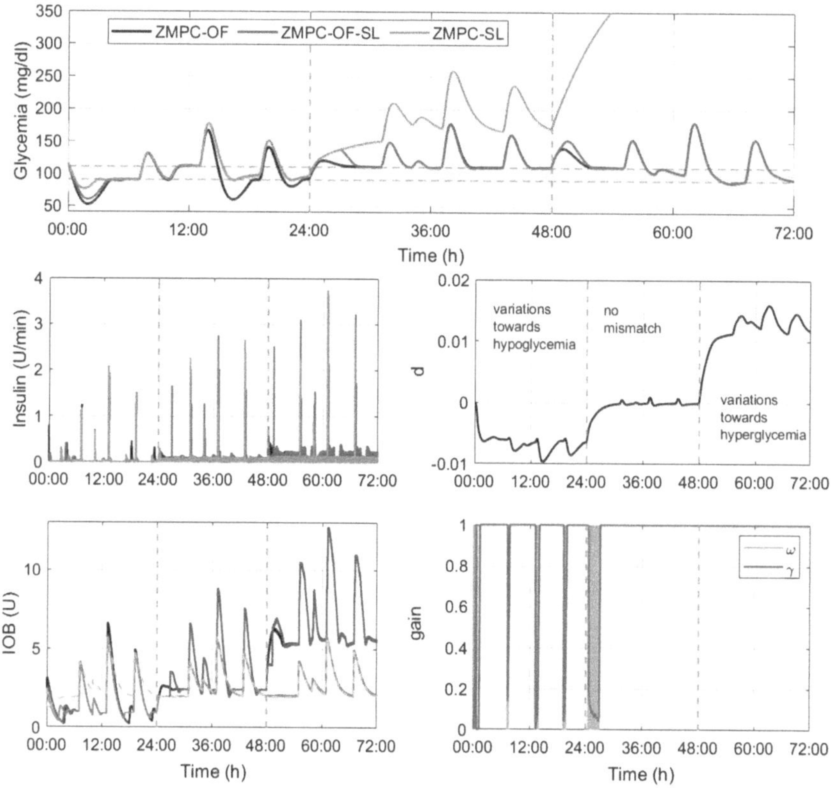

FIGURE 4.4 System evolution of the virtual adult subject number 10 when applying the ZMPC-OF and different ways of coupling the SL.

In this work, the information of the mismatch is used to reset the SL boundary, tightening or releasing it, to anticipate metabolic variations leading to hypoglycemia or hyperglycemia, respectively. A simple rule was then established to reset the boundary of the open-loop IOB profile. When $d < -\epsilon$ a tightened constraint is set with $\beta = 0.6$, when $d \in [-\epsilon, \epsilon]$ a normal constraint is set as $\beta = 1$, and the constraint is released when $d > \epsilon$, i.e., $u_f = u$. The ϵ value was empirically set to 0.002 for all virtual subjects.

In the same Fig. 4.4, two uses of the SL are depicted: (i) the red (gray in print version) lines show the ZMPC-OF strategy coupled with the SL while a factor $\beta = 0.6$ is maintained during the whole simulation to constrain the IOB (this strategy is denoted as ZMPC-SL), and (ii) the blue (dark gray in print version) lines show the ZMPC-OF strategy with the SL when the information of d is used to adapt the IOB boundary by changing β (this strategy is denoted as ZMPC-OF-SL).

It is observed that in fact, on day 1, the SL is beneficial with either case, (i) or (ii), because it achieves its goal of eliminating hypoglycemic events in the postmeal periods. From the IOB plot, it is visualized that on this first day the ZMPC-SL and the ZMPC-OF-SL keep insulin restricted in the safe IOB zone delimited with a factor of $\beta = 0.6$. Hence, the insulin doses are reduced with respect to the ZMPC-OF, and as a consequence BG levels are better regulated. Nevertheless, from days 2 and 3, it is observed that the ZMPC-SL is detrimental by maintaining a strong restriction on the IOB when it is not necessary. This prevents the controller from supplying the insulin required to decrease BG levels, being very damaging on the third day where the physiological variations that induce hyperglycemia occur.

In contrast, by adapting the boundary $\overline{IOB}$, the ZMPC-OF-SL achieves to maintain an adequate control on days 2 and 3 since the boundary $\overline{IOB}$ changes according to the d value. When hypoglycemic variations are detected, the $\overline{IOB}$ is reset using a factor $\beta = 0.6$ (see day 1), when d values near zero are detected, higher insulin dosage is allowed by changing the factor to $\beta = 1$ (see day 2), and when hyperglycemia-inducing variations are detected, the constraint is released and thus the behavior of the ZMPC-OF is emulated. This analysis shows the benefits of using the information of the plant–model mismatch to use some additional tools (as the IOB) when it is required.

Finally, the ZMPC-OF-SL strategy was evaluated using the 3-day simulation scenario and compared with the ZMPC-OF for the complete cohort of 10 virtual adult patients as seen in Fig. 4.5. Dotted black lines represent the population median when applying the ZMPC-OF, and solid red (gray in print version) lines represent the median with the ZMPC-OF-SL. Shaded areas are the interquartile range of the population. Despite a common fall at the beginning of day 1, due to the initial convergence of the disturbance estimator, the benefit of pairing the SL with the disturbance estimator is clear. The approach achieved to avoid hypoglycemic episodes in the postmeal periods and managed to keep near normoglycemic values in the consequent challenges. This is also supported by the outcome metrics reported in Table 4.3, where the performance of the strategies is compared (ZMPC-OF vs. ZMPC-OF-SL). The ZMPC-OF-SL achieves improved control performance in terms of time percentage in normoglycemia [70–180 mg/dl] (90.4 ± 6.8 vs. 94.5 ± 3.9); the time percentage in hypoglycemia is reduced (5.8 [1.6] vs. 2.2 ± 1.5) without incurring a higher percentage in hyperglycemia.

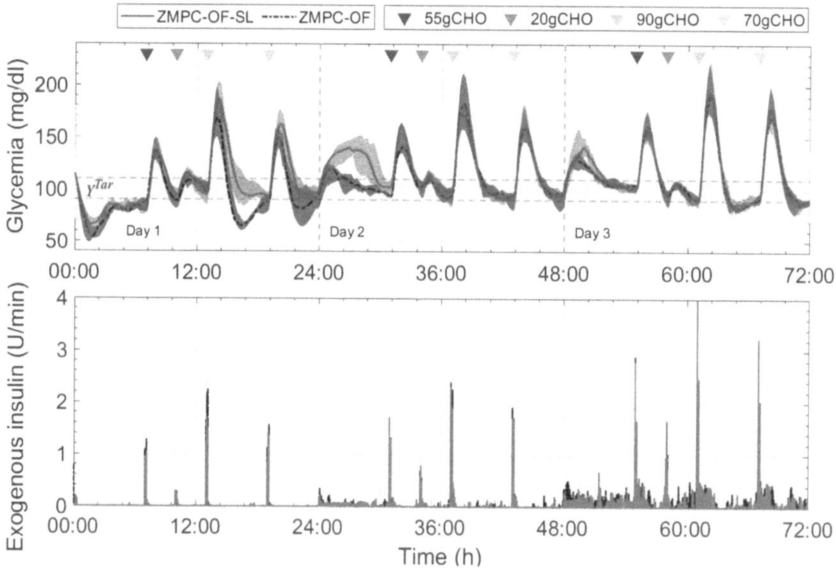

FIGURE 4.5 Glycemia evolution and exogenous insulin in 10 virtual adult patients when applying control strategies with pulse inputs and there is a plant–model mismatch. Solid (ZMPC-OF-SL) and dotted (ZMPC-OF) lines represent the median, shaded areas represent the interquartile range.

TABLE 4.3 Performance comparison of the ZMPC-OF and the ZMPC-OF-SL.

Strategy	**ZMPC-OF**	**ZMPC-OF-SL**
Mean BG (mg/dl)	108.0 (2.1)	113.4 ± 6.3
SD BG (mg/dl)	28.2 ± 5.9	28.5 ± 6.6
CV BG (mg/dl)	26.4 ± 5.1	25.1 ± 5.0
Time percentage of BG in each range (%)		
<54 mg/dl	1.2 ± 1.8	0 (0.5)
<70 mg/dl	5.8 (1.6)	2.2 ± 1.5
70–140 mg/dl	80.1 ± 6.6	79.9 ± 7.4
70–180 mg/dl	90.4 ± 6.8	94.5 ± 3.9
>180 mg/dl	1.6 (5.3)	1.6 (6.4)
>250 mg/dl	0 (0)	0 (0)
Number of events in each range (-)		
<54 mg/dl	1 (1)	0 (0)
<70 mg/dl	2.5 (1)	1.4 ± 1.3
>180 mg/dl	1.5 (4)	2.8 ± 2.7
>250 mg/dl	0 (0)	0 (0)

4.6 Conclusion

The problem of designing an offset-free MPC with pulse inputs to compensate parameter variations that cause a plant–model mismatch has been addressed. The strategy consists of augmenting the system with a disturbance model to estimate the mismatch and correct the prediction model, equilibrium, and target in the MPC formulation. In addition, the pulse control scheme is based on strategies whose inputs act for a certain time without fully covering the sampling period (as in discrete schemes). This allows more realistic models for applications such as the artificial pancreas. The offset-free control scheme is applied to a cohort of 10 virtual subjects for T1DM treatment and the performance is compared with a ZMPC formulation developed in a previous work.

In addition, the problem of hypoglycemia events generated by insulin stacking was addressed. To that purpose, a SL based on the IOB and coupled to the ZMPC-OF was presented. Simple rules to tighten or release the SL were implemented according to the estimation of the plant–model mismatch d, reflecting the variations towards hypo- and hyperglycemia. The ZMPC-OF strategy with and without the SL was tested under a 3-day simulation scenario with variations in the plant that induce hypoglycemia and hyperglycemia and sensor noise. Satisfactory results were obtained by reducing hypoglycemic events without dangerously increasing BG levels in an adult cohort.

References

[1] P.S. Rivadeneira, A. Ferramosca, A.H. González, Control strategies for nonzero set-point regulation of linear impulsive systems, IEEE Trans. Autom. Control 63 (2018) 2994–3001, https://doi.org/10.1109/TAC.2017.2776598.

[2] P.S. Rivadeneira, C.H. Moog, Observability criteria for impulsive control systems with applications to biomedical engineering processes, Automatica 55 (2015) 125–131, https://doi.org/10.1016/j.automatica.2015.02.042.

[3] P.S. Rivadeneira, A. Ferramosca, A.H. Gonzalez, Impulsive zone model predictive control with application to type I diabetic patients, in: 2016 IEEE Conf. Control Appl. CCA 2016, 2016, pp. 544–549.

[4] A.J.N. Anelone, M.F. Villa-Tamayo, P.S. Rivadeneira, Oncolytic virus therapy benefits from control theory, R. Soc. Open Sci. 7 (2020) 200473, https://doi.org/10.1098/rsos.200473.

[5] P.S. Rivadeneira, C.H. Moog, Impulsive control of single-input nonlinear systems with application to HIV dynamics, Appl. Math. Comput. 218 (2012) 8462–8474, https://doi.org/10.1016/j.amc.2012.01.071.

[6] A. Haidar, The artificial pancreas: how closed-loop control is revolutionizing diabetes, IEEE Control Syst. 36 (2016) 28–47, https://doi.org/10.1109/MCS.2016.2584318.

[7] R. Gondhalekar, E. Dassau, F.J. Doyle, Periodic zone-MPC with asymmetric costs for outpatient-ready safety of an artificial pancreas to treat type 1 diabetes, Automatica 71 (2016) 237–246, https://doi.org/10.1016/j.automatica.2016.04.015.

[8] P. Abuin, P.S. Rivadeneira, A. Ferramosca, A.H. González, Artificial pancreas under stable pulsatile MPC: improving the closed-loop performance, J. Process Control 92 (2020) 246–260, https://doi.org/10.1016/j.jprocont.2020.06.009.

[9] P.S. Rivadeneira, A.H. González, Non-zero set-point affine feedback control of impulsive systems with application to biomedical processes, Int. J. Syst. Sci. 49 (2018) 3082–3093, https://doi.org/10.1080/00207721.2018.1533049.

[10] P.S. Rivadeneira, A. Ferramosca, A.H. González, MPC with state window target control in linear impulsive systems, IFAC-PapersOnLine 48 (2015) 507–512, https://doi.org/10.1016/j.ifacol.2015.11.329.

[11] P. Sopasakis, P. Patrinos, H. Sarimveis, A. Bemporad, Model predictive control for linear impulsive systems, IEEE Trans. Autom. Control 60 (2015) 2277–2282, https://doi.org/10.1109/TAC.2014.2380672.

[12] M.F. Villa Tamayo, M.A. Caicedo Alvarez, P.S. Rivadeneira, Handling parameter variations during the treatment of type 1 diabetes mellitus: in silico results, Math. Probl. Eng. 2019 (2019) 1–21, https://doi.org/10.1155/2019/2640405.

[13] M.F. Villa-Tamayo, M.A. Caicedo, P.S. Rivadeneira, Offset-free MPC strategy for nonzero regulation of linear impulsive systems, ISA Trans. 101 (2020) 91–101, https://doi.org/10.1016/j.isatra.2020.01.005.

[14] M. Morari, U. Maeder, Linear offset-free model predictive control, Automatica 48 (2012) 2059–2067, https://doi.org/10.1016/j.automatica.2012.06.038.

[15] G. Pannocchia, J.B. Rawlings, Disturbance models for offset-free model-predictive control, AIChE J. 49 (2003) 426–437, https://doi.org/10.1002/aic.690490213.

[16] G. Pannocchia, M. Gabiccini, A. Artoni, Offset-free MPC explained: novelties, subtleties, and applications, IFAC-PapersOnLine 48 (2015) 342–351, https://doi.org/10.1016/j.ifacol.2015.11.304.

[17] A. El Fathi, V. Gingras, B. Boulet, The artificial pancreas and meal control, IEEE Control Syst. (2018).

[18] P.S. Rivadeneira, M. Caicedo, A. Ferramosca, A.H. Gonzalez, Impulsive zone model predictive control (iZMPC) for therapeutic treatments: application to HIV dynamics, in: 2017 IEEE 56th Annu. Conf. Decis. Control, Vol. 2018- Janua, IEEE, 2017, pp. 4094–4099.

[19] M.F. Villa-Tamayo, A.J.N. Anelone, P.S. Rivadeneira, Tumor reduction using oncolytic viruses under an impulsive nonlinear estimation and predictive control scheme, IEEE Control Syst. Lett. 5 (2021) 1705–1710, https://doi.org/10.1109/LCSYS.2020.3043185.

[20] D. Mayne, J. Rawlings, Constrained model predictive control: stability and optimality, Automatica 36 (2000) 789–814, https://doi.org/10.1016/s0005-1098(00)00173-4.

[21] E. Fushimi, N. Rosales, H. De Battista, F. Garelli, Artificial pancreas clinical trials: moving towards closed-loop control using insulin-on-board constraints, Biomed. Signal Process. Control 45 (2018) 1–9, https://doi.org/10.1016/j.bspc.2018.05.009.

[22] F. León-Vargas, F. Garelli, H. De Battista, J. Vehí, Postprandial blood glucose control using a hybrid adaptive PD controller with insulin-on-board limitation, Biomed. Signal Process. Control 8 (2013) 724–732, https://doi.org/10.1016/j.bspc.2013.06.008.

[23] C. Ellingsen, E. Dassau, H. Zisser, B. Grosman, M.W. Percival, L. Jovanovič, et al., Safety constraints in an artificial pancreatic β cell: an implementation of model predictive control with insulin on board, J. Diabetes Sci. Technol. 3 (2009) 536–544, https://doi.org/10.1177/193229680900300319.

[24] M. Messori, G.P. Incremona, C. Cobelli, L. Magni, Individualized model predictive control for the artificial pancreas, IEEE Control Syst. Mag. 38 (2018) 86–104, https://doi.org/10.1109/MCS.2017.2766314.

[25] C. Dalla Man, F. Micheletto, D. Lv, M. Breton, B. Kovatchev, C. Cobelli, The UVA/PADOVA type 1 diabetes simulator: new features, J. Diabetes Sci. Technol. 8 (2014) 26–34, https://doi.org/10.1177/1932296813514502.

[26] Y. Ruan, M.E. Wilinska, H. Thabit, R. Hovorka, Modeling day-to-day variability of glucose-insulin regulation over 12-week home use of closed-loop insulin delivery, IEEE Trans. Biomed. Eng. 64 (2017) 1412–1419, https://doi.org/10.1109/TBME.2016.2590498.

[27] J.D. Hoyos, M.F. Villa-Tamayo, C.E. Builes-Montano, A. Ramirez-Rincon, J.L. Godoy, J. Garcia-Tirado, et al., Identifiability of control-oriented glucose-insulin linear models: review and analysis, IEEE Access 9 (2021) 69173–69188, https://doi.org/10.1109/ACCESS.2021.3076405.

[28] T. Danne, R. Nimri, T. Battelino, R.M. Bergenstal, K.L. Close, J.H. DeVries, et al., International consensus on use of continuous glucose monitoring, Diabetes Care 40 (2017) 1631–1640, https://doi.org/10.2337/dc17-1600.

[29] M.F. Villa-Tamayo, F. Leon-Vargas, M. Garcia-Jaramillo, P.S. Rivadeneira, Glycemic control strategy based on an impulsive MPC with safety layer coupling for IOB limitation, IEEE Control Syst. Lett. 5 (2021) 1669–1674, https://doi.org/10.1109/LCSYS.2020.3042651.

[30] N. Magdelaine, P.S. Rivadeneira, L. Chaillous, A-L. Fournier-Guilloux, M. Krempf, T. MohammadRidha, et al., Hypoglycaemia-free artificial pancreas project, IET Syst. Biol. 14 (2019) 16–23, https://doi.org/10.1049/iet-syb.2018.5069.

[31] M.F. Villa-Tamayo, P.S. Rivadeneira, Adaptive impulsive offset-free MPC to handle parameter variations for type 1 diabetes treatment, Ind. Eng. Chem. Res. 59 (2020) 5865–5876, https://doi.org/10.1021/acs.iecr.9b05979.

CHAPTER

5

Deep neuronal network-based glucose prediction for personalized medicine

Alma Y. Alanis , Oscar D. Sanchez , and Eduardo Ruiz Velázquez

University Center for Exact Sciences and Engineering, University of Guadalajara, Guadalajara, Jalisco, Mexico

5.1 Introduction

Diabetes mellitus (DM) involves a set of metabolic disorders, including carbohydrate, fat, and protein metabolism disorders. The number of patients with DM in 2014 was estimated to be approximately 422 million adults worldwide [58]. The main characteristic is that glucose levels can rise abnormally to ranges above 300 mg/dL; this event is called hyperglycemia and occurs after food intake (postprandial). This is caused by a defect in secretion, action, or both of the hormone insulin. Insulin plays an important role in the control of glucose levels in the blood; in fact, metabolic disorders in carbohydrates, lipids, and proteins are attributed to problems in the assimilation or secretion of insulin, which highlights the importance of this anabolic hormone. Low supply of insulin to regulate glucose levels in addition to insulin resistance developed by tissues, muscle, adipose tissue, the liver, and effector enzymes or genes are some of the factors responsible for metabolic disorders [28].

Uncontrolled DM care can lead to complications such as nerve and brain damage, heart diseases, vision loss, amputations, kidney diseases, and in more severe cases death [41], [14]. In fact, according to the World Health Organization, there were 1.5 million deaths that were directly attributed to DM in 2012 [58].

Feedback Control for Personalized Medicine
https://doi.org/10.1016/B978-0-32-390171-0.00014-7

According to the classification of DM, a different approach is taken for its treatment. In general, as proposed by the American Diabetes Association (ADA), it is classified into four types: type 1, type 2, other types, and gestational DM [5].

Patients with DM are mostly diagnosed in two of the etiopathogenetic categories. The cause of type 1 DM (T1DM) is insufficient secretion of insulin by the β-cells of the pancreas. People at risk of developing T1DM can be identified by serological evidence, in which anti-β-cell antibodies (ICAs), anti-insulin antibodies (IAAs), and anti-glutamic acid decarboxylase (GAD) antibodies are detected as a result of an autoimmune process. These antibodies are of predictive interest because they usually appear on occasions before the onset of this disease [5].

Type 2 DM is a disease characterized by its common development in adulthood. It is the result of a combination of resistance to the action of insulin, which causes an increase in the need for insulin in the target tissues. Therefore, the β-cells of the pancreas cannot meet the high insulin demand because they are defective. The majority of patients with type 2 DM do not show dependence on insulin; in this case insulin secretion continues and insulin depletion does not occur frequently.

The main difference between type 1 and type 2 DM is insulin dependence. Another characteristic difference is the absence of ketoacidosis in the majority of patients with type 2 DM, and also the autoimmune destruction of the pancreatic β-cells does not occur [28].

In individuals with type 2 DM, the pancreas is capable of producing insulin, at least initially, and sometimes for life, but the tissues are resistant to insulin. Patients diagnosed with this type of diabetes do not need external insulin to survive. In fact, some patients with type 2 DM maintain their glucose levels within normal ranges through weight reduction, exercise, and/or oral hypoglycemic agents [5].

On the other hand, current treatment for patients diagnosed with T1DM is based on periodic blood glucose measurements and insulin injections to regulate glucose values. In both type 1 and type 2 DM, maintaining the glucose level between 70 and 180 mg/dL is not an easy task due to alterations such as food intake, stress, exercise, and the complexity of glucose metabolism [56], [1].

However, patients on exogenous insulin treatment who deliver a higher dose than needed may experience hypoglycemic episodes (glucose levels below 70 mg/dL). Hyperglycemic and hypoglycemic events cause macrovascular damage, such as peripheral, cerebral, and coronary vascular insufficiency. They are also associated with microvascular diseases such as neuropathy, retinopathy, and nephropathy. If these conditions persist for a long period of time they cause increased disability, reduced life expectancy, and costly healthcare for families and governments [6].

Tight glucose control would reduce serious complications related to the disease. With this purpose, emerging technologies have developed increasingly accurate sensors for continuous glucose monitoring (CGM), in the same way also in continuous subcutaneous insulin infusion devices. The availability of these devices allows the realization of sensor-augmented pump (SAP) therapy [46], which helps the patient to maintain the glucose concentration in the glycemic range. However, the SAP is not autonomous, because the patient needs to calculate the amount of insulin that has to be delivered by the pump based on the measurements made by the glucose sensor.

An autonomous biomedical insulin infusion device to carry out glucose level control in patients with T1DM can be achieved by the artificial pancreas (AP) [20], [55]. Currently the PA systems are portable and less invasive, providing a high degree of freedom. In general, the AP architecture consists of a subcutaneous insulin pump for the administration of insulin, a CGM for glucose measurement (sensor), and a control algorithm for the execution of the control.

One of the main components of the AP is precisely the control algorithm, which calculates the amount of insulin to supply subcutaneously during food intake and postprandial periods. Recently, in the work published in [42], the AP was used continuously for a month, 24 hours a day. The AP was controlled by a predictive control algorithm model. In the work carried out by [34], an individualized model predictive control algorithm was proposed based on the idea that the glucose–insulin dynamics of patients differs from each other.

Glucose–insulin models in the context of AP are so important that the reliability of the model's glucose predictions directly affects the performance of the controller. In fact, compartmental models of insulin glucose [26], [12], simulators [32], and neural network (NN) models [17], [60], [15], [23], [35], [45], among others, have been published [57], [21], [18].

In general, glucose–insulin models have been reported as a one-step forward predictor with satisfactory results [3]. However, the information provided by a one-step forward predictor may be insufficient, and it is sometimes desirable to obtain information at various steps in the future.

Multistep ahead (MSA) prediction refers to predicting several values in sequence in the future using only the values observed in the past, resulting in a complex task even using NNs. MSA prediction has been a topic of interest in several scientific studies [9], [10]. MSA prediction with NN has been used successfully in [52], [11]. For T1DM treatment, MSA prediction has been a topic of interest because it is important to avoid risk events (hypoglycemia and hyperglycemia); it is undoubtedly very beneficial to know the future glucose conditions.

Glucose sensors allow continuous glucose measurement. In this way, a time series method could be implemented to solve prediction problems focused on individual patients. In [27] and [33], it was proposed to incorporate an alarm in the CMG in real time when glucose rises above the hyperglycemic threshold or falls below the hypoglycemic threshold, which means an important improvement in glucose therapy [8], [19]. That is, an alert generated 20 to 30 min before a risk event occurs could provide sufficient information for diabetes care [61]. Therefore, a methodology capable of preventing future risk events is necessary [61]. Furthermore, this knowledge of future glucose concentrations over a short or long period of time could help a control algorithm improve its performance to prevent hyperglycemic or hypoglycemic conditions.

Currently, some sensors use simple methods to generate an alert when the sensor considers that the trend in glucose concentration will cross the threshold. Among these we can mention those based on a polynomial [51], AR/ARMA [22], state space [38], autoregression (AR), AR exogenous input (ARX), AR moving average exogenous input (ARMAX) [21], and NN models [40], [39]. In the same manner, they are proposed for short-term glucose prediction [7], [50], and weighted recursive least squares is used for online identification [19]. According to published T1DM prediction results, a prediction horizon of 30–45 min is sufficient.

However, MSA prediction is a difficult task to develop because it is not possible to determine closely related input information in time to adjust the performance of the model. In addition to this, a small one-step prediction error accumulates and spreads into the future. In this work, we propose the use of deep NNs (DNNs) to perform online MSA prediction using as input real data provided by a CGM. All NN models are trained to directly predict the output with a prediction horizon of 15, 30, 45, and 60 min. The results show that the prediction precision of NNs is satisfactory. Therefore, the proposed DNNs are a reliable method of MSA prediction of glucose concentrations, which can be implemented online based on CGM information.

This work is organized as follows. First, the NNs used are shown, followed by a description of the direct prediction strategy and the structure of the system. Experimental results are shown including a comparison between the networks, and finally the conclusions are stated.

5.2 Deep neural networks

DNNs are called "deep" because they are constructed by several layers of nonlinear operations, each one next to the other, in such a way that they can abstract representations of the real world with their weights (parameters) [37]. DNNs have been shown to be very effective for complicated real-world problems such as classification, image processing [29], speech processing [48], language models [59], and sequential data [53], [25].

The dataset considered in this work will be described in more detail later. In general, these data are sequential in nature. Therefore, we propose

NN architectures that are adept in learning nonlinearity in time series data, in order to capture the behavior of a person with T1DM to perform MSA prediction. We consider a feedforward multilayer network and recurrent networks with long short-term memory (LSTM) (unidirectional and bidirectional).

5.2.1 Recurrent neuronal networks

Recurrent NNs (RNNs) provide a way to deal with time sequenced data correlated between data points nearby in the sequence [31]. RNNs have feedforward connections, to themselves and to other cells of the previous layers [44]. The RNN depends on the state of the network at the previous time in addition to external inputs to produce the output.

Let $x_1, ..., x_k$ represent a sequence of length k and let y be the output vector of each element of the sequence of x by computing the following equations from $t = 1$ to k:

$$h_t = H(W_{xh}x_t + W_{hh}h_{t-1} + b_h), \tag{5.1}$$

$$y_t = W_{hy}h_t + b_y, \tag{5.2}$$

where h_t are the hidden nodes, Ws are weight matrices (W_{xh} are the weights that connect the input layer and the hidden layer, W_{hh} connect the hidden layer and the previous hidden states, and W_{hy} connect the hidden layer and the output), b is the bias vector, and H is the activation function of the hidden layer. Eq. (5.1) represents the connection between the previous and the current hidden states.

5.2.2 Long short-term memory recurrent neural network

LSTM is an RNN network variant and is a popular RNN structure in the field of deep learning, which is able to address the vanishing gradient. It was proposed by Hochreiter and Schmidhuber in 1997 [25]. RNNs have difficulties in learning long-term dependencies, but the LSTM network solves the long-term dependency problem by introducing three gated units: input and output gates and a forget gate; with these three gates the memory of past states is controlled. LSTM has been applied in time series modeling and sequence prediction with satisfactory results [24]. Among them we can mention for the prediction of time series the works carried out by [43], [4], [62], a model has been reported to predict blood glucose levels one step ahead with the LSTM in [16]. LSTM can learn to solve complex tasks [25].

The structure of the LSTM cell is shown in Fig. 5.1 [24], where i is the input gate, f is the forget gate, and o is the output gate.

An LSTM is implemented by the following functions [49]:

$$i_t = \sigma(W_{xi}x_t + W_{hi}h_{t-1} + W_{ci}C_{t-1} + b_i), \tag{5.3}$$

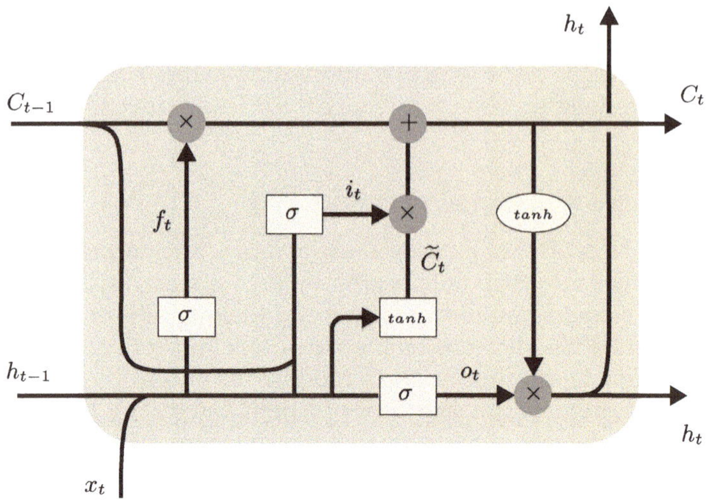

FIGURE 5.1 Extended LSTM cell that introduces a peephole connection. Peephole connections allow control the memory cell of the LSTM.

$$f_t = \sigma(W_{xf}x_t + W_{hf}h_{t-1} + W_{cf}C_{t-1} + b_f), \tag{5.4}$$

$$\widetilde{C}_t = tanh(W_{xc}x_t + W_{hc}h_{t-1} + b_c), \tag{5.5}$$

$$C_t = f_t C_{t-1} + i_t \widetilde{C}_t, \tag{5.6}$$

$$o_t = \sigma(W_{xo}x_t + W_{ho}h_{t-1} + W_{co}C_t + b_o), \tag{5.7}$$

$$h_t = o_t \tanh(C_t), \tag{5.8}$$

where C_t is the memory cell, $\widetilde{C}_t$ represents the new candidate values that will be added into the LSTM memory, σ is defined as the logistic sigmoid function, and h_t is its representation as a value between -1 and 1; b are learned biases, the Ws are weight matrices, and tanh scales the values into the range -1 to 1.

5.2.3 Bidirectional LSTM

Schuster and Paliwal in 1997 introduced for the first time the bidirectional recurrent NNs (BRNNs) [47]. The aim is to divide the state of the neurons of the RNN, one responsible for the forward state (positive direction) and the other for the backward state (negative direction). The outputs of the forward state and the inputs of the backward are independent of each other [47]. This bidirectional structure is applied to RNN variants; here, we use bidirectional LSTM. Incorporating a bidirectional structure, each LSTM cell is enabled to access the state from the two directions (past and future).

The two hidden layers are $\overrightarrow{h_t}$, which computes the forward sequence, and $\overleftarrow{h_t}$, which computes the backward sequence. Then, $\overrightarrow{h_t}$ and $\overleftarrow{h_t}$ are combined to produce the output y_t. The hidden state h is an LSTM cell. The Bi-LSTM is implemented by the following equations:

$$\overrightarrow{h_t} = H(W_{x\overrightarrow{h}} x_t + W_{\overrightarrow{h}\overrightarrow{h}} \overrightarrow{h}_{t-1} + b_{\overrightarrow{h}}), \tag{5.9}$$

$$\overleftarrow{h_t} = H(W_{x\overleftarrow{h}} x_t + W_{\overleftarrow{h}\overleftarrow{h}} \overleftarrow{h}_{t-1} + b_{\overleftarrow{h}}), \tag{5.10}$$

$$y_t = W_{\overrightarrow{h}y} \overrightarrow{h}_t + W_{\overleftarrow{h}y} \overleftarrow{h}_t + b_y. \tag{5.11}$$

The bidirectional structure is shown in Fig. 5.2.

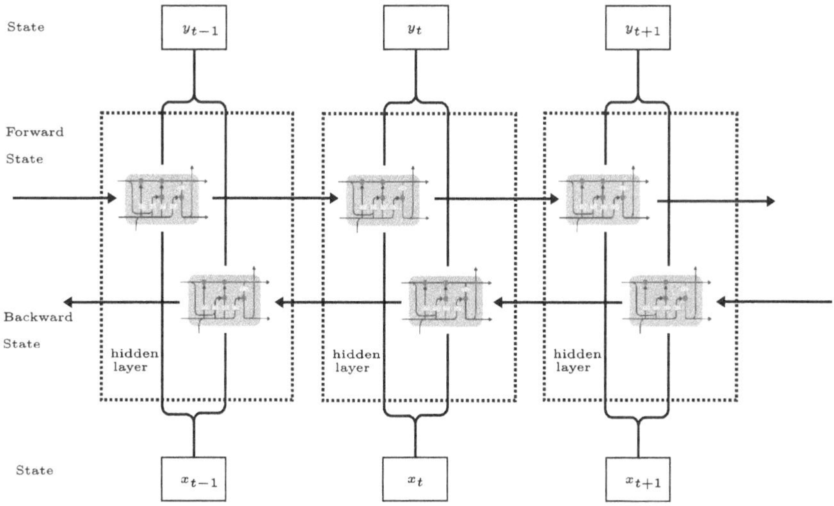

FIGURE 5.2 Representative diagram of the structure of a bidirectional recurrent neural network.

The bidirectional LSTM receives training for both the forward and backward state and since there is no interaction between both state neurons it can be trained with the same algorithms as a unidirectional LSTM.

5.2.4 Multilayer networks

The multilayer perceptron (MLP) NN is one of the most influential network models and is widely used in various problems. The structure of the MLP network consists of several nodes organized in layers, where the nodes are connected with nodes of other adjacent layers by weights. The MLP consists of an input layer, one or more hidden layers, and an output layer.

The first layer is calculated by means of the sum of the n neurons in the hidden layer, obtained by multiplying the connection weights w_{nn} and its input nodes u_n, then it is multiplied by the weights w_{nm} of the second layer, the sum is evaluated in the activation function to generate the output y_p.

$$y_p = \varphi(\sum w_{nm}\varphi(\sum w_{nn}u_n)), \tag{5.12}$$

where φ is the activation function. Fig. 5.3 represents the MLP NN architecture used for prediction.

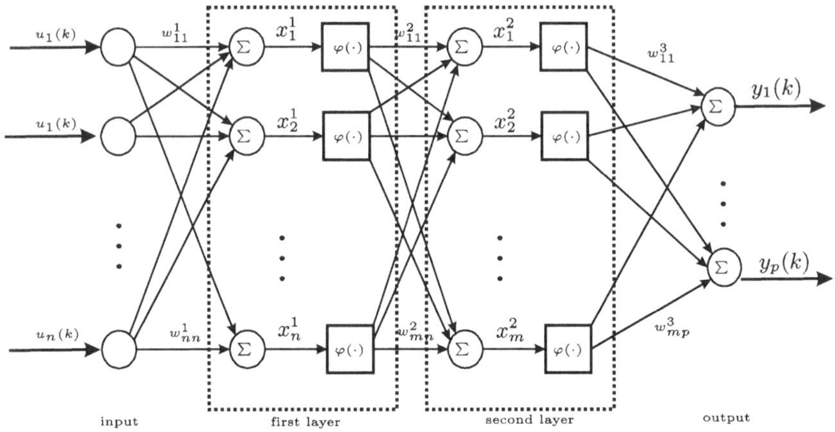

FIGURE 5.3 MLP neural network structure with two hidden layers.

5.3 Direct multistep ahead prediction strategy

A multistep prediction consists of predicting the value $[\hat{y}(k+n)]$ of the time series $[y(1), \ldots, y(k)]$ with k observations, where n denotes the prediction horizon; if $n = 1$ it is called short term, for $n > 1$ it is called long term.

There are two common strategies for MSA forecasts: (1) recursive prediction, also called iterative, and (2) direct prediction. In MSA forecasting, for the recursive method the model is trained to predict one step forward, and then using the same model the predicted value is added to the previous values in the input to predict the next step. On the other hand, for the second method, the model is trained so that it generates the direct prediction at time $k+n$ [10], [11].

Multiple studies have compared the different strategies; however, it is not clear which strategy is better. Regardless of which method is used, both strategies present the problem of propagation of the error for a long prediction horizon, which causes the precision of the predictions to decrease.

The direct strategy (also called independent) [54] consists of predicting n steps forward independently of the others. That is, n models f_i are needed to predict each horizon of the time series $[y(1), ..., y(k)]$, where

$$y(t+i) = f_i(y(t), ..., y(t-d+1)) + e \tag{5.13}$$

with $t \in \{d, ..., k-n\}$ and $i \in \{1, ..., n\}$, where f is the functional dependency between the observations, d is the number of delays used to predict, and e is the error, disturbances, and/or noise.

The prediction is calculated using the n trained models $\hat{f}_i$ as follows:

$$\hat{y}(k+i) = \hat{f}_i(y(k), ..., y(k-d+1)) + e, \tag{5.14}$$

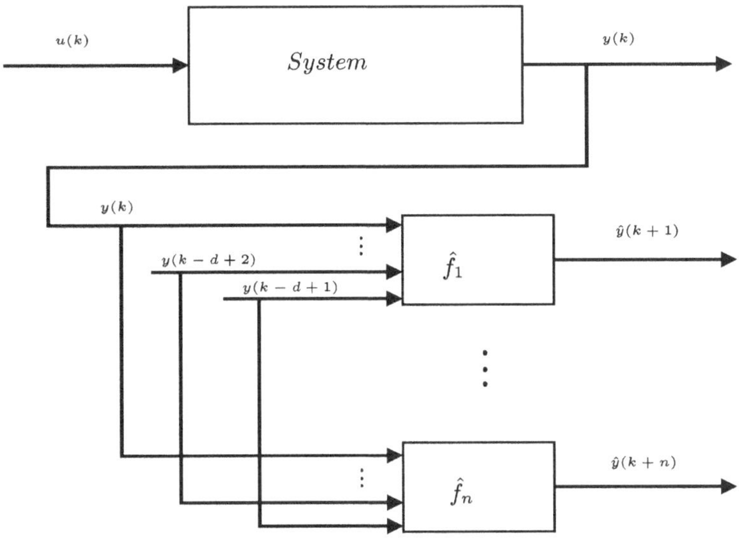

FIGURE 5.4 Direct MSA prediction architecture.

The n models learn independently, so there is independence between the n predictions; also, the direct strategy does not use estimate values to predict, so it does not accumulate too much error [54]. The architecture of the direct strategy of MSA prediction is shown in Fig. 5.4.

5.4 System description

The information used in this work was acquired from the Minimed Real-Time Paradigm. This system consists mainly of two devices, an insulin pump and a CGM. This system is in an open circuit, since the amount of carbohydrates in the food is calculated using the instructions of an insulin pump guide. The amount of insulin that the pump delivers to the patient is then programmed based on their current condition. In addition, there is a glucose sensor that provides the levels (every 5 min) of glucose in interstitial blood for CGM.

The user can access the data provided by the glucose sensor after the sensor is disconnected from the patient's body. These data can be downloaded to a computer for further analysis [2].

5.4.1 Dataset description

The experimental dataset is downloaded directly from the aforementioned CGM device. This corresponds to a 23-year-old female patient, 14 year old T1DM diagnosis, with a height of 1.68 m and a weight of 58.5 kg. In general, this information can be separated into three parts, defined as: (1) the input signal of carbohydrate intake, representing the disturbances (u_c), (2) the input of subcutaneous insulin, considered as the control signal (u_i), and (3) the output signal, the blood glucose level (BGL) measured by the sensor due to the ingestion and administration of insulin (y). These data correspond to 7500 min, which corresponds to just over 5 days that the CMG was connected to the patient.

A standard diet was followed under medical supervision corresponding to three meals a day with snacks. The patient leads a normal life so the patient may forget to follow the diet or provide this information to the insulin pump.

The amounts of food that the patient ingested (orange (light gray in print version) line) and the insulin boluses administered to the patient (green (mid gray in print version) line) are shown in Fig. 5.5; these boluses programmed by the patient always occur at the same time: the intake. In addition, the average basal insulin level is shown: 1.1 mU/h.

Fig. 5.6 shows the interstitial glucose level measured by the glucose sensor (red (gray in print version) line). As reported in [30], it must be considered that there is a difference of 17.2 ± 7.2 min between the appearance of glucose in the blood and the interstitial glucose. This is because when food is ingested, glucose does not appear instantly in the blood; it takes approximately 2 hours to appear in the blood after ingestion. This delay is not considered in the insulin pump, and therefore it would be of great help to a glucose controller to improve its performance. The data presented in Fig. 5.6 were used to make the prediction.

5.4.2 Neuronal network configuration using CGM

NNs have been trained to predict glucose levels 15, 30, and 45 min ahead to compare the performance for the three proposed networks. That

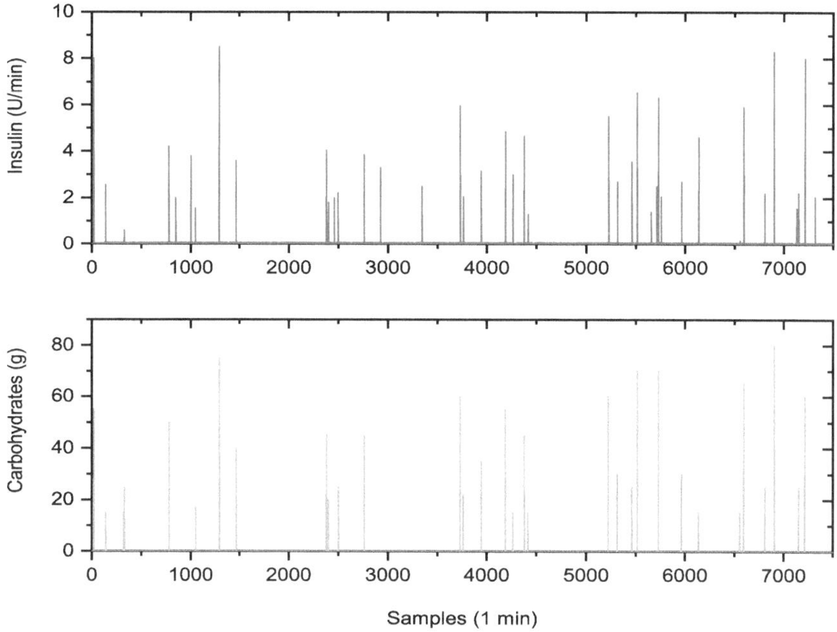

FIGURE 5.5 The insulin (U/min) and the carbohydrate intake (g) for the patient.

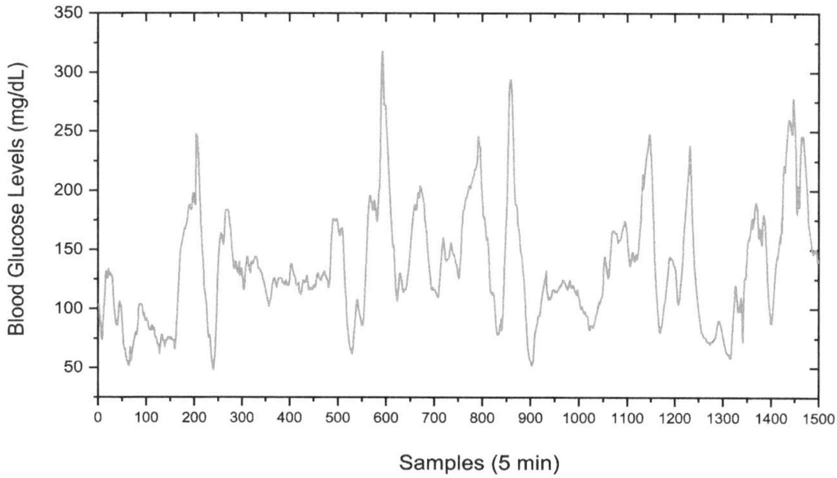

FIGURE 5.6 Glucose levels obtained by the CMG (mg/dL).

is to say, the NNs are trained to generate a direct prediction of $y(k+n)$, where n corresponds to the required prediction time.

The information that enters the NNs is the current glucose measurement $y(k)$, in addition to some past glucose samples measured by the CGM system. To produce the prediction $\hat{y}(k+n)$, the three NN models take into account past glucose information up to 40 min before the current time.

Fig. 5.7 shows the NN configuration using CGM data. Since the sensor records glucose levels every 5 min, the NN model needs eight samples from the sensor as inputs, equivalent to a memory of 40 min. The NNs were trained using only data from the sensor as inputs; 60% of the data was used for training and 40% was used for validation.

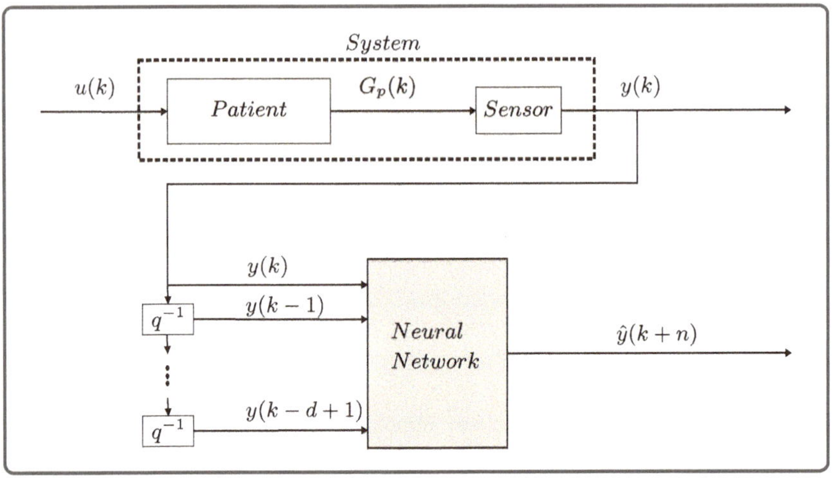

FIGURE 5.7 Neuronal network configuration using CGM data. $G_p(k)$ represents the patient's blood glucose level in the sample k.

5.5 Results

After trying different configurations for the networks, we obtained the best results with the following structures described for each of the networks used. It is worth mentioning that the same structure was used for the different prediction horizons. The LSTM NN used a structure with two hidden layers with 100 and 50 neurons, respectively. Similarly, the BiLSTM NN has two hidden layers, with 100 neurons for the first layer and 50 for the second. Finally, the MLP NN has a structure of two hidden layers, with 20 neurons for the first layer and 15 for the second layer. The weights and bias of the NNs were randomly initialized, and training was carried out by the Levenberg–Marquardt backpropagation algorithm for MLP and by the Adam algorithm for LSTM and BiLSTM. The training is applied in batches.

The prediction results for 15, 30, 45, and 60 min forward for each of the networks are shown in Figs. 5.8, 5.9, and 5.10. The results show that the precision of the three NN predictions are satisfactory.

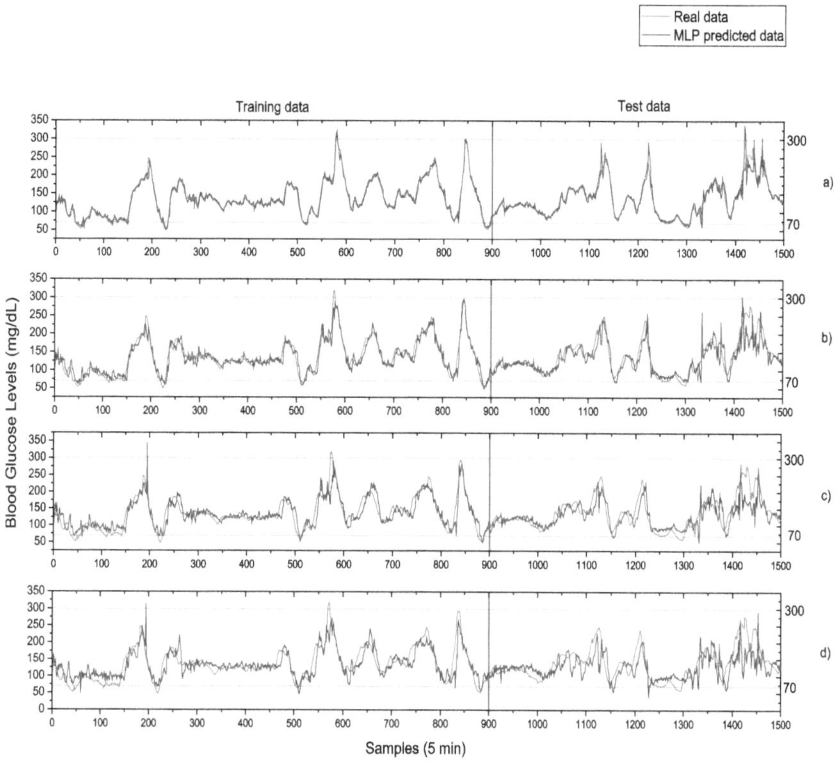

FIGURE 5.8 MLP neuronal network prediction results for (a) 15, (b) 30, (c) 45, and (d) 60 min forward.

The NN prediction for 15 min forward presents little error. This glucose would roughly mean the delay that exists between blood glucose and interstitial glucose. With a prediction of 15 min ahead, the blood glucose would be estimated approximately, which would allow the control algorithms to calculate the insulin dose with higher pressure. Despite this, the prediction time is quite short, so there is not enough time to correct the glucose trend either by the patient or by a control algorithm. Graphically, it is observed that the results for the three algorithms are apparently similar, so there is not much difference between them.

The results shown by the three NNs for the prediction from 30–45 min forward are quite good, and predicting the patient's blood glucose at this time could help him/her to know the current condition, so the patient could anticipate severe cases of risk conditions. Finally, the results of a prediction of up to 60 min forward showed that the three proposed NNs begin to present a larger error, being more evident in the LSTM and BiLSTM networks.

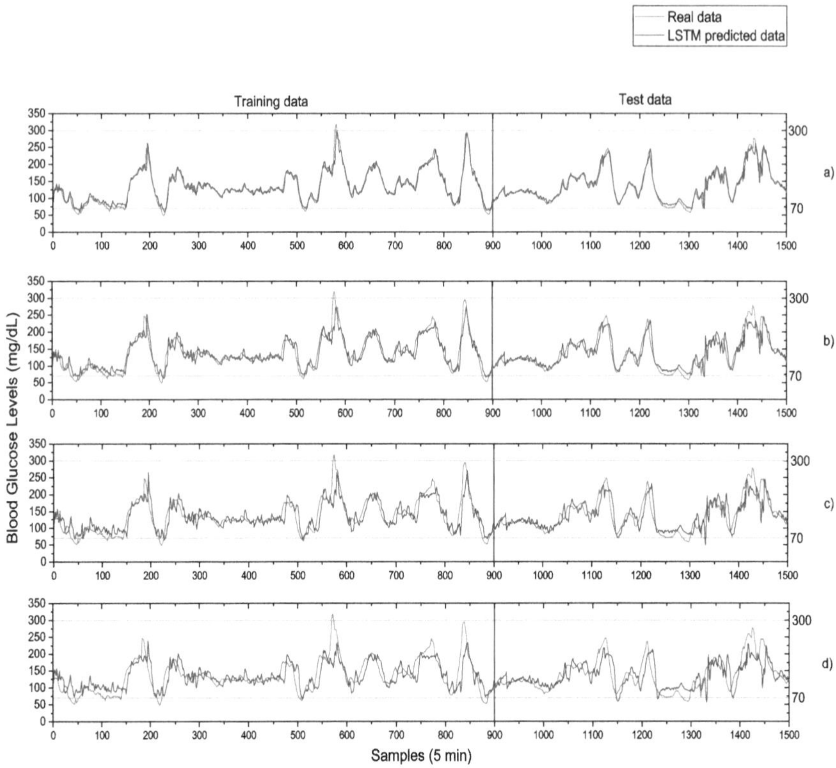

FIGURE 5.9 LSTM neuronal network prediction results for (a) 15, (b) 30, (c) 45, and (d) 60 min forward.

The usefulness of the 30–45 min or longer prediction lies in the fact that a glucose monitor could know in advance when glucose levels cross the risk boundaries (300 mg/dL and 70 mg/dL). In this way the controller can make decisions according to the risk condition for the patient, so that the time in which hyperglycemia occurs can be minimized or at least reduced by avoiding administering insulin when the glucose trend is about to cross the 70 mg/dL limit; on the other hand, with knowledge of the increasing trend of the patient's glucose, the controller can deliver more insulin if required to prevent hypoglycemia.

Since the performance of the three NNs presents similar results, the models are evaluated using the evaluation criteria that indicate the general predictive ability by comparing the real datasets and the predicted data by the NN. The evaluation criteria used were: correlation coefficient, root mean square error (RMSE [mg/dL]), and time lag (min).

RMSE indicates the difference between the real data y and the predicted data $\hat{y}$. A lower RMSE means better prediction performance. It is calcu-

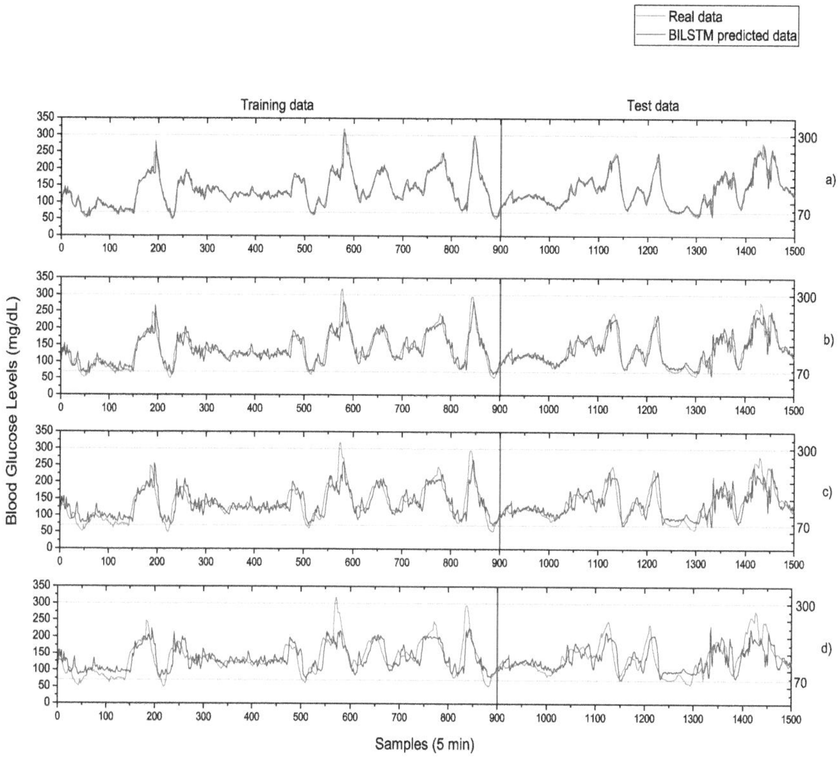

FIGURE 5.10 BiLSTM neuronal network prediction results for (a) 15, (b) 30, (c) 45, and (d) 60 min forward.

lated using the following equation:

$$RMSE = \sqrt{\sum(\hat{y} - y)^2}. \tag{5.15}$$

For this work the Pearson and Spearman correlation coefficients were used to know the linear dependence between real data and predicted data [36]:

Pearson's correlation coefficient is calculated as follows:

$$CC_p = \frac{\sum_{i=1}^{N}(y_i - y_{mean})(\hat{y}_i - \hat{y}_{mean})}{\sqrt{\left[\sum_{i=1}^{N}(y_i - y_{mean})^2\right]\left[\sum_{i=1}^{N}(\hat{y}_i - \hat{y}_{mean})^2\right]}}, \tag{5.16}$$

where y_i and $\hat{y}_i$ are values of y and $\hat{y}$ for the ith individual.

Spearman's correlation coefficient is calculated as follows:

$$CC_s = 1 - \frac{6\sum_{i=1}^{N} d_i^2}{N(N^2-1)}, \tag{5.17}$$

where d_i is the difference in ranks for y and $\hat{y}$.

The results of the correlation coefficient for Spearman and Pearson can be interpreted from Table 5.1.

TABLE 5.1 Rule for interpreting correlation coefficients [36].

Size	Interpretation
0.90 to 1.00	Very high correlation
0.70 to 0.90	High correlation
0.50 to 0.70	Moderate correlation
0.30 to 0.50	Low correlation
0.00 to 0.30	Negligible correlation

In Figs. 5.8, 5.9, and 5.10 it can be seen that the predictions are not lagged in time by the same extent as the prediction horizon n. The time delay can be obtained by shifting the predictions against the data, calculating the RMSE in each offset to find that offset in time that minimizes the RMSE [13].

Graphically the prediction curve slides along the time axis until it finds the best match with the actual data curve. This time-lagged function is defined as follows:

$$T_\tau = \sqrt{\frac{1}{N-\tau}\sum_{i=1}^{N-\tau}(\hat{y}_i - y_{i+\tau})^2}, \tag{5.18}$$

where τ is the time shift. For this work τ is considered in multiples of 5 min, because we only have knowledge of blood glucose levels every 5 min.

Table 5.2 presents the results of the different evaluation criteria of all the networks. It can be seen that for a prediction horizon greater than 30 min, the BiLSTM NN achieves the best results for the RMSE; the best correlation coefficient for the prediction horizon of 30–45 min is presented by the LSTM and BiLSTM; and MLP is the method that presents the least delay in time, so in this way it performs better in MSA prediction.

On the other hand, the LSTM and BiLSTM NNs do not show much difference in their performance; both are slightly better for the prediction horizon of 15 min.

TABLE 5.2 Prediction results of different methods.

Prediction	EC	LSTM	BiLSTM	MLP
15 min ahead	RMSE	11.4216	11.0353	12.1274
	SD	7.8205	7.4163	7.4091
	CC_p	0.9733	0.9740	0.9686
	CC_s	0.97810	0.97842	0.97934
	Time lag	8	8	7
30 min ahead	RMSE	19.4782	19.4697	19.6444
	SD	14.0660	13.9622	13.7486
	CC_p	0.9169	0.9160	0.9155
	CC_s	0.92337	0.92287	0.92256
	Time lag	18	18	13
45 min ahead	RMSE	26.6837	26.6150	26.6362
	SD	19.4513	19.4998	19.2005
	CC_p	0.8353	0.8363	0.8371
	CC_s	0.84318	0.84174	0.84325
	Time lag	31	28	23
60 min ahead	RMSE	32.4991	32.0818	32.1084
	SD	24.1588	24.0195	23.6945
	CC_p	0.7430	0.7508	0.7550
	CC_s	0.7525	0.75211	0.75346
	Time lag	38	38	26

In Fig. 5.11 a comparative image between the three networks is shown. For this graph, only the sample interval from 800 to 1000 was taken into account to observe in better detail the prediction of the NNs.

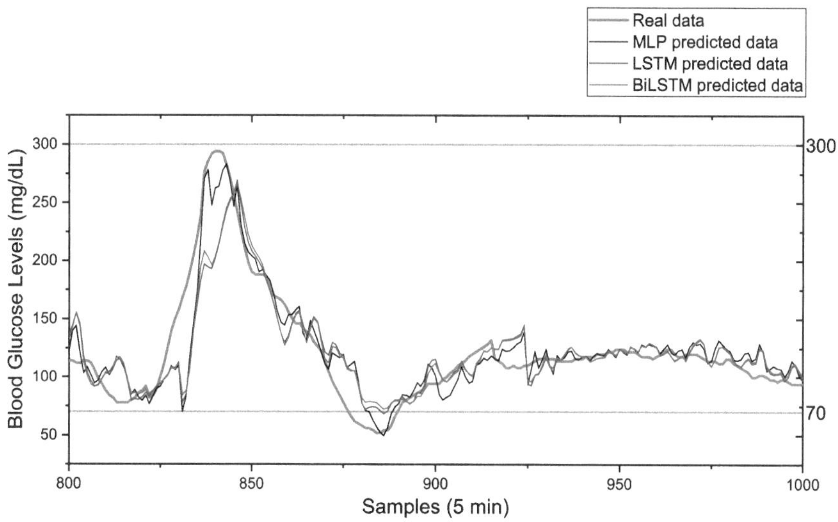

FIGURE 5.11 Comparative 45 min ahead prediction results of different methods.

5.6 Conclusion and discussion

In this work, three DNNs were used to perform the MSA prediction of a patient with T1DM. Considering the behavior of the nonlinear glucose dynamics, the results were satisfactory for all networks. It was found that the NN that has a better time lag performance for MSA prediction was the MLP NN, this being the one with the lowest time shift, and LSTM and BiLSTM have a higher correlation coefficient. Overall, DNNs have proven to be a very powerful tool for MSA prediction, and are an alternative to improving the processing capabilities by incorporating the prediction in the CMG device. It should be noted that most glucose prediction works use the inputs of food intake and insulin supply by the pump; however, this study considers that they are inputs that cannot be accessed so easily and that it would need the cooperation of the patient to have access to this information, to maintain in this way the autonomy of the glucose treatment devices. These inputs are not necessary and only consider the sensor output without filtering it.

The direct strategy for prediction requires offline training, which allows the proposed NNs to adjust its parameters to maintain a close approximation to the measurement, in addition to maintaining a good prediction for a long time of data with a low RMSE, independently of glucose dynamics that vary throughout the day due to external alterations such as exercise, stress, or diet. In this way, these networks could be trained for individual patients according to their particular conditions. In addition, they can be trained again after obtaining more samples from the sensor and since they maintain an acceptable prediction for several hours, the NN can be updated every period of time to maintain these results.

An automatic control scheme with integration of predictions several steps in the future using NNs could lead to exclusive systems for a single patient rather than a general population. Therapeutically it can significantly improve the treatment of this disease. Although the predictor error 30–45 min forward has an RMSE between 19.4782 and 26.6837 (mg/dL), it should be noted that the intention of knowing the future trend of glucose is to anticipate possible scenarios of the patient of hyperglycemia and hypoglycemia, to prevent them.

Acknowledgments

The authors thank the University of Guadalajara for giving us the support to develop this research.

References

[1] R. Aathira, V. Jain, Advances in management of type 1 diabetes mellitus, World Journal of Diabetes 5 (2014) 689, https://doi.org/10.4239/wjd.v5.i5.689.

[2] A.Y. Alanis, B.S. Leon, E.N. Sanchez, E. Ruiz-Velazquez, Blood glucose level neural model for type 1 diabetes mellitus patients, International Journal of Neural Systems 21 (2011) 491–504, https://doi.org/10.1142/S0129065711003000.
[3] A.Y. Alanis, E.N. Sanchez, A.G. Loukianov, Discrete-time adaptive backstepping nonlinear control via high-order neural networks, IEEE Transactions on Neural Networks 18 (2007) 1185–1195, https://doi.org/10.1109/TNN.2007.899170.
[4] F. Altché, A. de La Fortelle, An LSTM network for highway trajectory prediction, in: 2017 IEEE 20th International Conference on Intelligent Transportation Systems (ITSC), IEEE, 2017, pp. 353–359.
[5] A.D. Association, et al., Diagnosis and classification of diabetes mellitus, Diabetes Care 37 (2014) S81–S90, https://doi.org/10.2337/dc14-S081.
[6] A.D. Association, et al., Standards of medical care in diabetes—2017 abridged for primary care providers, Clinical Diabetes 35 (2017) 5, https://doi.org/10.2337/cd16-0067.
[7] B.W. Bequette, Continuous glucose monitoring: real-time algorithms for calibration, filtering, and alarms, Journal of Diabetes Science and Technology 4 (2010) 404–418, https://doi.org/10.1177/193229681000400222.
[8] T. Bremer, D.A. Gough, Is blood glucose predictable from previous values? A solicitation for data, Diabetes 48 (1999) 445–451, https://doi.org/10.2337/diabetes.48.3.445.
[9] F. Chee, T. Fernando, Closed loop system to automate insulin infusion. In closed-loop control of blood glucose, IEEE Transactions on Information Technology in Biomedicine 157 (2007).
[10] P.A. Chen, L.C. Chang, F.J. Chang, Reinforced recurrent neural networks for multi-step-ahead flood forecasts, Journal of Hydrology 497 (2013) 71–79, https://doi.org/10.1016/j.jhydrol.2013.05.038.
[11] J. Chi, H.c. Kim, Prediction of Arctic sea ice concentration using a fully data driven deep neural network, Remote Sensing 9 (2017) 1305, https://doi.org/10.3390/rs9121305.
[12] C. Cobelli, C. Dalla Man, G. Sparacino, L. Magni, G. De Nicolao, B.P. Kovatchev, Diabetes: models, signals, and control, IEEE Reviews in Biomedical Engineering 2 (2009) 54–96, https://doi.org/10.1109/RBME.2009.2036073.
[13] A.J. Conway, K.P. Macpherson, J.C. Brown, Delayed time series predictions with neural networks, Neurocomputing 18 (1998) 81–89, https://doi.org/10.1016/S0925-2312(97)00070-2.
[14] S.R. Dikondwar, Design and development of insulin delivery system prototype, 2011, pp. 575–579.
[15] A.H. El-Baz, A.E. Hassanien, G. Schaefer, Identification of Diabetes Disease Using Committees of Neural Network-Based Classifiers, Springer, 2016.
[16] T. El Idriss, A. Idri, I. Abnane, Z. Bakkoury, Predicting blood glucose using an lstm neural network, in: 2019 Federated Conference on Computer Science and Information Systems (FedCSIS), IEEE, 2019, pp. 35–41.
[17] A.K. El-Jabali, Neural network modeling and control of type 1 diabetes mellitus, Bioprocess and Biosystems Engineering 27 (2005) 75–79, https://doi.org/10.1007/s00449-004-0363-3.
[18] M. Eren-Oruklu, A. Cinar, L. Quinn, D. Smith, Adaptive control strategy for regulation of blood glucose levels in patients with type 1 diabetes, Journal of Process Control 19 (2009) 1333–1346, https://doi.org/10.1016/j.jprocont.2009.04.004.
[19] M. Eren-Oruklu, A. Cinar, D.K. Rollins, L. Quinn, Adaptive system identification for estimating future glucose concentrations and hypoglycemia alarms, Automatica 48 (2012) 1892–1897, https://doi.org/10.1016/j.automatica.2012.05.076.
[20] R. Femat, E. Ruiz-Velázquez, G. Quiroz, Weighting restriction for intravenous insulin delivery on t1dm patient via h_∞ control, IEEE Transactions on Automation Science and Engineering 6 (2009) 239–247, https://doi.org/10.1109/TASE.2008.2009089.
[21] D.A. Finan, C.C. Palerm, F.J. Doyle, H. Zisser, L. Jovanovic, W.C. Bevier, D.E. Seborg, Identification of empirical dynamic models from type 1 diabetes subject data, 2008.

[22] A. Gani, A.V. Gribok, S. Rajaraman, W.K. Ward, J. Reifman, Predicting subcutaneous glucose concentration in humans: data-driven glucose modeling, IEEE Transactions on Biomedical Engineering 56 (2008) 246–254, https://doi.org/10.1109/TBME.2008.2005937.

[23] M.A. González-Olvera, A.G. Gallardo-Hernández, Y. Tang, M.C. Revilla-Monsalve, S. Islas-Andrade, A discrete-time recurrent neurofuzzy network for black-box modeling of insulin dynamics in diabetic type-1 patients, International Journal of Neural Systems 20 (2010) 149–158, https://doi.org/10.1142/S0129065710002322.

[24] A. Graves, A.r. Mohamed, G. Hinton, Speech recognition with deep recurrent neural networks, in: 2013 IEEE International Conference on Acoustics, Speech and Signal Processing, IEEE, 2013, pp. 6645–6649.

[25] S. Hochreiter, J. Schmidhuber, Long short-term memory, Neural Computation 9 (1997) 1735–1780, https://doi.org/10.1162/neco.1997.9.8.1735.

[26] R. Hovorka, F. Shojaee-Moradie, P.V. Carroll, L.J. Chassin, I.J. Gowrie, N.C. Jackson, R.S. Tudor, A.M. Umpleby, R.H. Jones, Partitioning glucose distribution/transport, disposal, and endogenous production during ivgtt, American Journal of Physiology: Endocrinology and Metabolism 282 (2002) E992–E1007, https://doi.org/10.1152/ajpendo.00304.2001.

[27] A. Kamath, A. Mahalingam, J. Brauker, Methods of Evaluating the Utility of Continuous Glucose Monitor Alerts, SAGE Publications, 2010.

[28] A.T. Kharroubi, H.M. Darwish, Diabetes mellitus: the epidemic of the century, World Journal of Diabetes 6 (2015) 850, https://doi.org/10.4239/wjd.v6.i6.850.

[29] A. Krizhevsky, I. Sutskever, G.E. Hinton, Imagenet classification with deep convolutional neural networks, Communications of the ACM 60 (2017) 84–90, https://doi.org/10.1145/3065386.

[30] E. Kulcu, J.A. Tamada, G. Reach, R.O. Potts, M.J. Lesho, Physiological differences between interstitial glucose and blood glucose measured in human subjects, Diabetes Care 26 (2003) 2405–2409.

[31] D.N. Kumar, K.S. Raju, T. Sathish, River flow forecasting using recurrent neural networks, Water Resources Management 18 (2004) 143–161, https://doi.org/10.1023/B:WARM.0000024727.94701.12.

[32] C.D. Man, F. Micheletto, D. Lv, M. Breton, B. Kovatchev, C. Cobelli, The Uva/Padova type 1 diabetes simulator: new features, Journal of Diabetes Science and Technology 8 (2014) 26–34, https://doi.org/10.1177/1932296813514502.

[33] G. McGarraugh, Alarm Characterization for a Continuous Glucose Monitor That Replaces Traditional Blood Glucose Monitoring, SAGE Publications, 2010.

[34] M. Messori, G.P. Incremona, C. Cobelli, L. Magni, Individualized model predictive control for the artificial pancreas: in silico evaluation of closed-loop glucose control, IEEE Control Systems Magazine 38 (2018) 86–104, https://doi.org/10.1109/MCS.2017.2766314.

[35] S. Mougiakakou, A. Prountzou, D. Iliopoulou, K. Nikita, A. Vazeou, C. Bartsocas, Nn based glucose-insulin metabolism models for children with type 1 diabetes, in: Conference Proceedings: Annual International Conference of the IEEE Engineering in Medicine and Biology Society. IEEE Engineering in Medicine and Biology Society. Conference, Vol. 1, 2006, pp. 3545–3548.

[36] M.M. Mukaka, A guide to appropriate use of correlation coefficient in medical research, Malawi Medical Journal 24 (2012) 69–71.

[37] O. Ogunmolu, X. Gu, S. Jiang, N. Gans, Nonlinear systems identification using deep dynamic neural networks, arXiv preprint, arXiv:1610.01439, 2016.

[38] C.C. Palerm, J.P. Willis, J. Desemone, B.W. Bequette, Hypoglycemia prediction and detection using optimal estimation, Diabetes Technology & Therapeutics 7 (2005) 3–14, https://doi.org/10.1089/dia.2005.7.3.

[39] S.M. Pappada, B.D. Cameron, P.M. Rosman, R.E. Bourey, T.J. Papadimos, W. Olorunto, M.J. Borst, Neural network-based real-time prediction of glucose in patients with insulin-dependent diabetes, Diabetes Technology & Therapeutics 13 (2011) 135–141, https://doi.org/10.1089/dia.2010.0104.

[40] C. Pérez-Gandía, A. Facchinetti, G. Sparacino, C. Cobelli, E. Gómez, M. Rigla, A. de Leiva, M. Hernando, Artificial neural network algorithm for online glucose prediction from continuous glucose monitoring, Diabetes Technology & Therapeutics 12 (2010) 81–88, https://doi.org/10.1089/dia.2009.0076.

[41] A. Pranckeviciene, J. Siudikiene, R. Ostrauskas, V. Machiulskiene, et al., Severity of periodontal disease in adult patients with diabetes mellitus in relation to the type of diabetes, Biomedical Papers of the Faculty of Medicine of Palacký University, Olomouc Czech Republic 158 (2014) 117–123.

[42] E. Renard, A. Farret, J. Kropff, D. Bruttomesso, M. Messori, J. Place, R. Visentin, R. Calore, C. Toffanin, F. Di Palma, et al., Day-and-night closed-loop glucose control in patients with type 1 diabetes under free-living conditions: results of a single-arm 1-month experience compared with a previously reported feasibility study of evening and night at home, Diabetes Care 39 (2016) 1151–1160, https://doi.org/10.2337/dc16-0008.

[43] M. Roondiwala, H. Patel, S. Varma, Predicting stock prices using lstm, International Journal of Science & Research 6 (2017) 1754–1756.

[44] C.O. Sakar, S.O. Polat, M. Katircioglu, Y. Kastro, Real-time prediction of online shoppers' purchasing intention using multilayer perceptron and LSTM recurrent neural networks, Neural Computing & Applications 31 (2019) 6893–6908, https://doi.org/10.1007/s00521-018-3523-0.

[45] W. Sandham, D. Hamilton, A. Japp, K. Patterson, Neural Network and Neuro-Fuzzy Systems for Improving Diabetes Therapy, Vol. 3, 1998, pp. 1438–1441.

[46] M. Schönauer, A. Thomas, Sensor-augmented pump therapy–on the way to artificial pancreas, Avances en Diabetología 26 (2010) 143–146, https://doi.org/10.1016/S1134-3230(10)63002-5.

[47] M. Schuster, K.K. Paliwal, Bidirectional recurrent neural networks, IEEE Transactions on Signal Processing 45 (1997) 2673–2681, https://doi.org/10.1109/78.650093.

[48] T.J. Sejnowski, C.R. Rosenberg, Parallel networks that learn to pronounce English text, Complex Systems 1 (1987) 145–168.

[49] S. Siami-Namini, N. Tavakoli, A.S. Namin, The performance of LSTM and BiLSTM in forecasting time series, in: 2019 IEEE International Conference on Big Data (Big Data), IEEE, 2019, pp. 3285–3292.

[50] G. Sparacino, A. Facchinetti, C. Cobelli, "Smart" continuous glucose monitoring sensors: on-line signal processing issues, Sensors (Basel, Switzerland) 10 (2010) 6751–6772, https://doi.org/10.3390/s100706751.

[51] G. Sparacino, F. Zanderigo, A. Maran, C. Cobelli, Continuous glucose monitoring and hypo/hyperglycaemia prediction, Diabetes Research and Clinical Practice 74 (2006) S160–S163, https://doi.org/10.2174/157339908785294361.

[52] H.T. Su, T.J. McAvoy, P. Werbos, Long-term predictions of chemical processes using recurrent neural networks: a parallel training approach, Industrial & Engineering Chemistry Research 31 (1992) 1338–1352, https://doi.org/10.1021/ie00005a014.

[53] I. Sutskever, Training Recurrent Neural Networks, University of Toronto, Toronto, Canada, 2013.

[54] S.B. Taieb, G. Bontempi, A.F. Atiya, A. Sorjamaa, A review and comparison of strategies for multi-step ahead time series forecasting based on the NN5 forecasting competition, Expert Systems with Applications 39 (2012) 7067–7083, https://doi.org/10.1016/j.eswa.2012.01.039.

[55] S. Trevitt, S. Simpson, A. Wood, Artificial pancreas device systems for the closed-loop control of type 1 diabetes: what systems are in development?, Journal of Diabetes Science and Technology 10 (2016) 714–723, https://doi.org/10.1177/1932296815617968.

[56] B. Tuch, M. Dunlop, J. Proietto, Diabetes Research: A Guide for Postgraduates, CRC Press, 2003.
[57] T. Van Herpe, M. Espinoza, B. Pluymers, I. Goethals, P. Wouters, G. Van den Berghe, B. De Moor, An adaptive input–output modeling approach for predicting the glycemia of critically ill patients, Physiological Measurement 27 (2006) 1057, https://doi.org/10.1088/0967-3334/27/11/001.
[58] W. WHO, Global report on diabetes, World Health Organization, 2016.
[59] W. Zaremba, I. Sutskever, O. Vinyals, Recurrent neural network regularization, arXiv preprint, arXiv:1409.2329, 2014.
[60] K. Zarkogianni, K. Mitsis, E. Litsa, M.T. Arredondo, G. Fico, A. Fioravanti, K.S. Nikita, Comparative assessment of glucose prediction models for patients with type 1 diabetes mellitus applying sensors for glucose and physical activity monitoring, Medical & Biological Engineering & Computing 53 (2015) 1333–1343, https://doi.org/10.1007/s11517-015-1320-9.
[61] C. Zecchin, A. Facchinetti, G. Sparacino, G. De Nicolao, C. Cobelli, Neural network incorporating meal information improves accuracy of short-time prediction of glucose concentration, IEEE Transactions on Biomedical Engineering 59 (2012) 1550–1560.
[62] Q. Zhang, H. Wang, J. Dong, G. Zhong, X. Sun, Prediction of sea surface temperature using long short-term memory, IEEE Geoscience and Remote Sensing Letters 14 (2017) 1745–1749, https://doi.org/10.1109/LGRS.2017.2733548.

CHAPTER

6

Control-based drug tailoring schemes towards personalized influenza treatment

Gustavo Hernandez-Mejia

Institute of Epidemiology and Social Medicine, University of Münster, Münster, Germany

6.1 Introduction

Influenza is an infectious respiratory disease caused by the influenza virus. There are four types of influenza viruses: A, B, C, and D. Influenza A virus (IAV) and influenza B virus spread and cause seasonal epidemics. The symptoms related to influenza infection vary from mild to severe, covering cough, sore throat, runny nose, headache, fever, fatigue, and muscle pain. The strength of the symptoms is usually associated with the strength of the infection, going from a moderate respiratory disease to severe and occasionally lethal pneumonia. Influenza virus-related diseases are more severe among high-risk groups such as toddlers under 59 months, pregnant women, seniors over 65 years old, immunocompromised individuals, and people with chronic conditions [1].

Diverse studies have tried to identify pandemics throughout the entire history, hence, reports of possible influenza pandemics can be found from the early Greek literature of 412 BCE to the more precise documentation of the present time. Along with several reports dating from the 14th and 15th centuries, the first reliable report was by Molineux in 1694 [2]. Also, diverse reports of influenza epidemics have been made for the 17th century in America and Europe. Since the 18th century, the quality and quantity of collected data increased and, for the first time, medical reports started to comment on the number of infected individuals, whether considering the event as an epidemic or a pandemic (see Definitions 6.1 and 6.2), the

Feedback Control for Personalized Medicine
https://doi.org/10.1016/B978-0-32-390171-0.00015-9

spread throughout countries, and the possible origins of the virus strains [3].

Definition 6.1. *Epidemic.* This refers to an increase, often sudden, in the number of cases of a disease above what is normally expected in the population of the area. Outbreak carries the same definition of epidemic but is often used for a more limited geographic area [4].

Definition 6.2. *Pandemic.* This indicates an epidemic that has spread over several countries and continents, usually affecting a massive number of people [4].

Four pandemics of human influenza have occurred in recent centuries. The 1918 pandemic caused by IAV subtype H1N1 is associated with more than 50 million deaths and hence considered the most devastating influenza pandemic to date [6]. As shown in Fig. 6.1, the influenza pandemics of 1957 and 1968 and the most recent one in 2009 were caused by IAV subtypes H2N2, H3N2, and H1N1, respectively. Human H3N2 viruses cocirculated with human 1918 lineage seasonal H1N1 viruses from 1977 to 2009, when this parentage was replaced by the new swine-origin H1N1 2009 pandemic virus due to reassortment. Since 2009, H3N2 viruses have cocirculated with 2009 pandemic lineage H1N1 viruses in humans. Seasonal influenza B viruses (not shown) are cocirculating in humans along with influenza A viruses [6,7]. In this sense, uncovering the secrets of the previous influenza pandemics and the virus involved has been extremely rewarding, but still, challenges remain a central necessity at diverse stages covering from epidemiology levels to host-related points, such as the immune response, virus pathogenesis, and treatment of influenza infection.

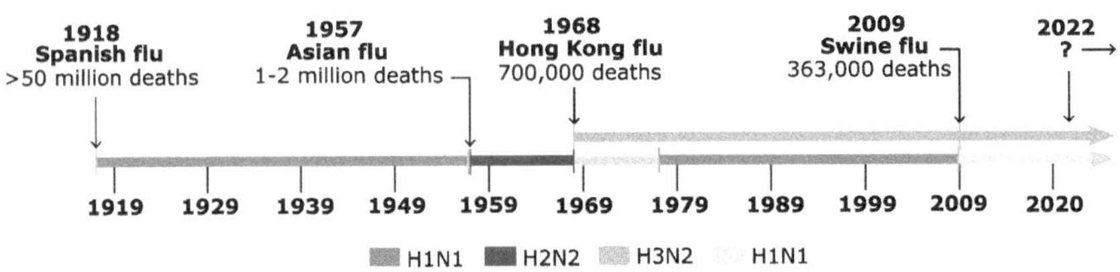

FIGURE 6.1 Influenza pandemics. After the 1918 Spanish flu pandemic, caused by a founder H1N1 influenza A virus, three following pandemics in 1957 (Asian flu), 1968 (Hong Kong flu), and 2009 (Swine flu) were caused by reassorted descendants of the 1918 virus. Color-coded timelines depict the years of annual epidemics of seasonal influenza that occurred after each pandemic along with the estimated number of fatalities. This shows that pandemic events including the circulating or emerging influenza strains may potentially occur in the next years. This figure was adapted from Taubenberger et al. [5].

6.2 Influenza virus in the host

6.2.1 Influenza virus structure

Influenza A and B viruses, which are enveloped negative-sense RNA viruses, have genomes containing eight single-stranded RNA segments inside the virus particle – the virion. The general structure is shown in Fig. 6.2. RNA segments encode the viral glycoproteins hemagglutinin (HA) and neuraminidase (NA). The HA glycoprotein has two characteristic structures: the "stalk" domain and the globular "head" domain. This protein mediates binding to sialic acid-containing receptors and viral entry. NA, on the other hand, allows releasing viruses bound to nonfunctional receptors, facilitating viral spread [7]. The RNA genome is bound by the viral nucleoprotein (NP) encoded by RNA segment 5, and the rest of the RNA segments, segments 7 and 8, encode the matrix protein (M1), the membrane protein (M2), and the nonstructural protein (NS). It is thought that the M1 protein provides a scaffold that helps the structure of the virion and regulates the trafficking of the viral RNA segments in the cell. The M2 protein is required for viral entry and exit and, together with HA and NA glycoproteins, is located on the surface of the virus anchored in a lipid membrane that comes from the infected cell [5].

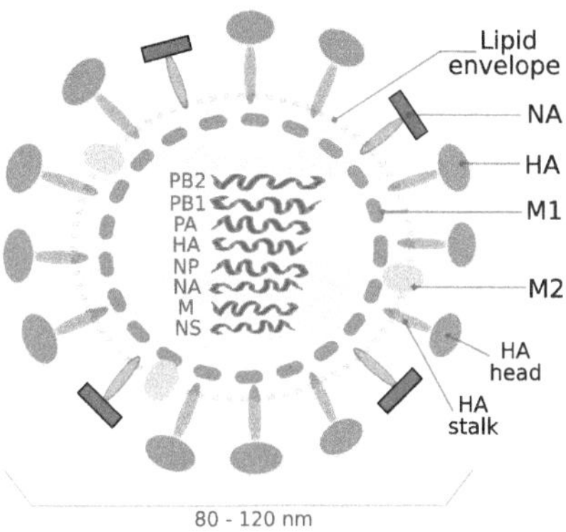

FIGURE 6.2 Influenza A and B virus structure. These viruses contain eight RNA segments, which encode RNA polymerase subunits, viral glycoproteins (hemagglutinin [HA], with its globular "head" and "stalk" structures, which facilitate viral entry, and neuraminidase [NA], which allows viral release), viral nucleoprotein (NP), a matrix protein (M1), a membrane protein (M2), and a nonstructural protein. The virion is formed by a lipid membrane derived from the infected cell. Figure adapted from Ref. [7] and Ref. [5].

The segmented nature of the influenza viral genome enables reassortment of genomic RNA segments, the so-called antigenic shift (Definition 6.3) when two viruses of the same type (two IAVs or two influenza B viruses) infect the same cell. Animal reservoirs as those of avian strains provide a source of antigenically diverse HA and NA genes that can be exchanged by reassortment after coinfection of the same host, increasing virus diversity and in some instances leading to the generation of human pandemic influenza virus strains with HA and NA derived from animal strains. Between pandemics, annual seasonal influenza A viruses are generated by continuous viral genetic mutations, antigenic drift (Definition 6.4), and intrasubtype reassortments [7–9].

Definition 6.3. *Antigenic shift.* The process by which two or more different strains of a virus, or strains of two viruses, combine to form a new subtype that has a mixture of the surface antigens of the two or more original strains. Antigenic shift is a specific case of reassortment or viral shift that confers a phenotypic change.

Definition 6.4. *Antigenic drift.* A type of genetic variation in viruses, arising by the accumulation of mutations in the virus genes that code for virus-surface proteins that host antibodies (Abs) recognize. The process results in a new strain of virions that are not effectively inhibited by Ab responses against previous strains.

The HA and NA viral proteins are the most antigenically variable and are even classified into antigenically diverse subtypes in the case of IAV. A total of 16 antigenically different HA and 9 antigenically different NA serotypes or subtypes have been identified among different avian strains [7]. Of note, since HA and NA viral glycoproteins are located at the surface of the virion, they are the main targets for protective Abs induced by influenza virus infection and vaccination [10,11].

6.2.2 Influenza virus pathogenesis

Influenza viruses are transmitted through different routes; in humans, for instance, through the respiratory route, whereas avian influenza viruses in wild birds are transmitted through the fecal–oral or fecal–respiratory routes [12,13]. The virus targets epithelial cells of the respiratory or intestinal tract for infection and productive replication. The severity of infection in humans is associated with replication of the virus in the lower respiratory tract, developing a serious inflammation process resulting from immune cell infiltration [14]. An IAV infection starts when virions enter the upper portion of the respiratory tract; once inside, they follow the different phases shown in Fig. 6.3. First, in the binding phase, virions attach to the receptors of the epithelial cells' surface due to the fusion capacity of the IAV HA protein [14]. Approximately 20 min after infection, the virion is endocytosed inside the epithelial cell (endocytosis phase) [15]. Next, in the cytoplasm, the viral RNA is released and used to produce new virions that are later released to the extracellular environment in the budding phase. The period between successful cell infection and virus release is an eclipse phase which ranges from 5 to 12 hours postinfection (hpi) [16]. The newly generated virions start to infect other cells favored by the NA activity, which prevents unfruitful binding of new virions' HA to receptors in the viral glycoproteins and on the membrane of the infected cells; this facilitates viral spread [17].

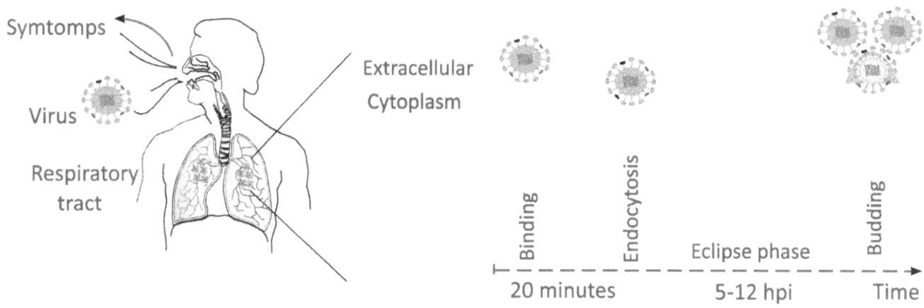

FIGURE 6.3 Influenza A virus infection phases. The main phases of IAV infection in the host. Virions enter the respiratory tract and bind to target cells. Afterward, virions enter the eclipse phase, which ranges 5–12 hours postinfection (hpi), and start to replicate before infecting other cells. This figure was adapted from Boianelli et al. [17].

6.2.3 Host immune response

Host immunity, composed of the innate and adaptive immune responses, represents a barrier that IAV needs to overcome to replicate and propagate in new hosts [18]. The innate immune system forms the first element of defense against infection, whose goal is to prevent respiratory epithelial cells from it, using mucus, collectins, and intracellular sensing mechanisms [19]. Signaling of various receptors (Toll-like receptors) leads to the production of proinflammatory cytokines and type I interferons (IFN-I), such as IFN-α and IFN-β, that inhibit protein synthesis in host cells [20,21]. After infection of the alveoli, alveolar macrophages become activated and phagocytose apoptotic IAV-infected cells, reducing viral spread [22]. Finally, natural killer (NK) cells recognize Ab-bound IAV-infected cells and lyse them through the Ab-dependent cell cytotoxicity process [18,23].

On the other hand, the adaptive immune response to influenza considers elements such as pathogen-specific Abs and cytotoxic CD8+ T cell responses [18]. Along with CD8+ T cells (cytotoxic T lymphocytes), regulatory T cells and CD4+ T cells display cytolytic activity to infected cells and they recognize and eliminate virus-infected cells, preventing the production of progeny virus [24]. Governed by B cell activity, the Ab response correlates with protective immunity and is considered an integral part of the response to both IAV infection and vaccination. During IAV pathogenesis, diverse proteins become reachable by Abs; HA and NA proteins are accessible on virions and infected cells, with a larger number of HA trimers than NA tetramers [25–27]. The responses of Abs that bind to HA and NA can be broadly protective. The study of the magnitude and breadth of Ab responses to divergent influenza strains is central to uncover insights on the effectiveness of vaccination strategies and to understand how immunity induced by one strain is effective against another, the so-called Ab cross-reaction [28–30].

6.3 Influenza treatment and therapy optimization

Along with vaccines, antiviral drugs are a key player in both the prevention and treatment of influenza virus infection and disease. For seasonal influenza, antivirals are mainly used for treating severely ill patients, particularly those with a compromised immune system. In a pandemic scheme, antiviral drugs are essential for both treating infected patients and preventing infection in exposed individuals. Currently, the World Health Organization (WHO) and the Food and Drug Administration (FDA) recommend the use of NA inhibitors (NIs) as the first-line treatment.

Definition 6.5. *Neuraminidase inhibitors (NIs).* A type of antiviral drug that disrupts the release of influenza virions from the host cell by mimicking the neuraminidase sialic acid binding site. NIs bind neuraminidase to the budding viruses, blocking their enzymatic activity.

Prior to 2004, the adamantane derivatives rimantadine and amantadine were drugs licensed for influenza treatment; however, from the 2004–2005 influenza season on, the use of adamantane derivatives was no longer recommended due to the emergence of resistance in most circulating influenza strains [31]. Oseltamivir, for instance, is an orally delivered NI prodrug, oseltamivir phosphate, which is afterward converted to its active carboxylate form in the liver. Zanamivir, on the other hand, is an inhaled NI powder; therefore, its administration route constantly limits the use in individuals with respiratory problems. A third NI drug is peramivir, which is intravenously administered and plays an option for hospitalized patients [32,33]. Study results report improved outcomes associated with NIs, including a reduced risk of pneumonia and reduced mortality in hospitalized patients. Better results have been observed with early NI administration, until 2 days after symptom onset; later administration can still be beneficial in severe cases [34,35].

Oseltamivir became the market leader due to its oral delivery being preferable over inhaled zanamivir. Nowadays, most of the developed countries primarily use NIs for the treatment of hospitalized or severely ill patients; however, the emergence of oseltamivir resistance is of particular concern for H1N1 viruses [36,37]. Other drugs than NIs that target different stages of the influenza virus life cycle in human cells have been developed and tested; among these are baloxavir marboxil (BAM) and favipiravir (FP) [38,39]. Further studies may focus on whether the enhanced reduction in viral replication will translate into reduced severity of disease in severely ill patients or if it may also reduce secondary transmission [40,41].

6.3.1 Optimizing influenza therapies

The treatment schedule of NIs, as recommended by FDA and WHO, usually consists of a fixed dose administered once or twice per day for a period of time [1,42]. However, the search for better medication strategies is an unceasing task that leads to a constant exploration of novel alternatives in drug development at diverse levels such as mechanisms of drug delivery and response, treatment scheduling optimization, and medication personalization. In this sense, the combination of different knowledge areas such as pharmacology, systems biology (SB), and control theory can guide the path to influenza treatment progress.

Definition 6.6. *PK/PD schemes.* Pharmacokinetics (PK) is frequently referred to as "*what the body does to the drug.*" This scheme includes the processes of absorption, distribution, metabolism, and excretion that govern the concentration time course of a drug in blood or plasma, using compartments as dynamics blocks of the concentration time course. Pharmacodynamics (PD), on the other hand, describes "*what the drug does to the body.*" This approach is centered on the drug effect depending on time and drug concentrations in the past – if available – until the present, along with additional covariates [43–45].

Regarding pharmacology, PK and PD modeling and simulation – the PK/PD approach (Definition 6.6) – has been extensively used in pharmaceutical companies and research groups that develop and apply mathematical and statistical methods to characterize, understand, and predict a drug's behavior [43]. Towards similar directions, SB provides system-level insights to understand interactions and dynamics at various scales, within cells, tissues, organs, and organisms, through mathematical and computational models. In recent years, modeling of host viral infections such as IAV has risen as a promising tool to quantify infection levels as well as to identify potential therapeutic targets for disease control [17]. Also recently, SB and pharmacology models – coupled as systems pharmacology – has guided research in precision medicine for complex diseases [46]. Lastly, control theory has resulted in widespread adoption of closed-loop (feedback) control in SB due to improvements in computing, actuation, and characterization of drugs' PK/PD properties, all using mathematical modeling platforms.

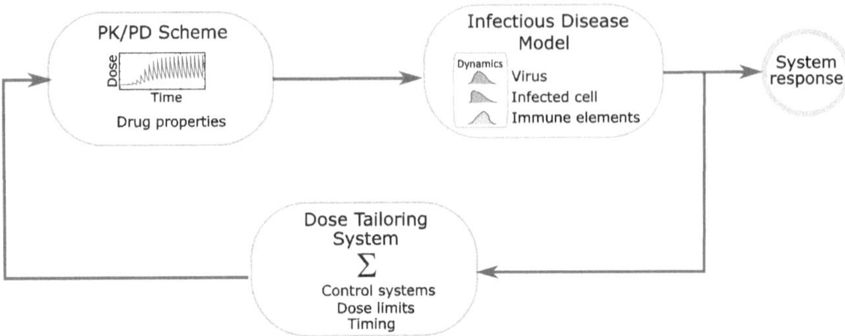

FIGURE 6.4 Control-based drug tailoring framework. The IAV dynamics, accounting for viral and host responses, are represented by a mathematical model that also reflects a clinical response. The dose tailoring system calculates the drug dose on the basis of IAV dynamics and the drug's properties, which follow a PK/PD model. This scheme forms a closed-loop system for automatically tailoring drug doses at a personal level.

Closed-loop drug estimation systems enable optimal scheduling of drug administration by leveraging classic and modern control-relevant models, systems pharmacology, and decision making, based on clinical targets [47]. Hence, influenza treatment scheduling is to be considered a control-based optimization duty, whose outcomes may guide the tailoring of treatment, in a framework of dose estimation. In this sense, the interactions between the virus, the immune response, infected and susceptible cells, and the drug's PK/PD dynamics, which may vary from one patient to another, are studied as a dynamical system to be analyzed by optimal control-based methods which also incorporate clinical requirements [48,49]. Thus, dose tailoring in the treatment scheduling framework is estimated depending on the evolution of each patient infection – a personalized scheme [48,47,49]. Fig. 6.4 depicts the structure of the control-based treatment tailoring. The first block of mathematical models describes biological phenomena such as infectious diseases, characterizing the viral and host dynamics, which also outline a clinical response. The second block, the dose tailoring system, computes the drug dose based on the viral, host, and drug responses. A closed-loop scheme is then formed adding a third block, the PK/PD models, standing for drug dynamics. Of note, since the control strategies perform based on the drug's PK/PD models, the complete approach can be also identified as a PK/PD-based control scheme [50].

6.4 Control schemes towards treatment tailoring

Theoretical control approaches can guide to design schedule treatments for infectious diseases such as influenza [48,51,52] and HIV [53,54,49]. Optimal control is related to finding a control law for a given system such that a performance criterion is minimized. This criterion is usually formulated as a cost functional, which is a function of state and control variables. While the optimal control allows determining a control law satisfying an optimality criterion, solving the associated Hamilton–Jacobi–Bellman (HJB) equation is rather a difficult assignment [55,56]. In this sense, to solve the optimal control problem for nonlinear systems with known mathematical models, diverse approaches are considered such as the dynamic programming methods [57], extended in the following years to obtain optimal control laws [58]. However, the ultimate scheme becomes an approximation dynamic programming (ADP) problem including iterative ADP algorithms and applications of ADP schemes [58–61].

An alternative strategy that supports optimality at the expense of a meaningful cost function is the inverse optimal control which avoids the solution of the HJB equation [62]. In the inverse fashion, a stabilizing feedback control law is established and then the corresponding cost functional is derived [63]. The inverse optimal control is a set of differential equations describing the shape of an objective function minimized with respect to the given control law and the initial state. In this case, for a stabilizing control law, one can find the objective function for a given stable dynamic system [64]. Hence, for the inverse approach, a stabilizing feedback control law based on prior knowledge of a control-Lyapunov function (CLF) can be designed first, and then it is established that this control law optimizes a cost functional; in fact, the main characteristic of the inverse approach is that the cost functional is a posteriori determined for the stabilizing feedback control law [64,65]. The inverse approach has been explored in the discrete-time framework, which presents desirable features for applicability in real-world scenarios [55,66–68].

Concerning robust feedback features, control via passivation supports obtaining a robust stabilizing feedback scheme [69,66]. Passivity systems theory is intimately linked to Lyapunov stability theory. There exist tools from the dissipative approach that can be used to generate Lyapunov functions. A difference between the two approaches is that the state of the system and the equilibrium point are notions required in the Lyapunov approach while the dissipative approach is rather based on the input–output behavior of the dynamic system. Considering these frameworks, a previous work [66] considers a quadratic storage function as a CLF to obtain discrete-time inverse optimal control laws. The existence of a CLF implies stability and this expression can be considered as a cost functional [55,70,71].

On the other hand, there is the need to characterize the dose intake in the control system, hypothesized as instantaneous incorporation of the drug into the system at a certain time instant – the drug intake. In this sense, impulsive control arises from the necessity to instantaneously change the state variables of a system. Impulsive control is, for instance, used if a system cannot be controlled by a continuous-time approach or if the input can only be applied at some time steps of a system, namely the control instants (τ_j) [72,73]. Control instants can be chosen whenever some conditions are satisfied, possibly generated by the control law. The control law $U(\tau_j, x)$ is somehow related to the concept of control input in other control methods; however, it refers to give a sudden change of state at instant τ_j. Diverse studies have recently explored applications and problem solutions using impulsive schemes, going from finite-time stability and optimization problems with impulsive triggers to control of disease [74–76]. Taking the machinery above together, this chapter integrates knowledge from the inverse optimal control theory, CLF stability and the impulsive control structure to design treatment schedules based on the viral dynamics [48,77] and the drug's PK/PD phases [50,78], and its potential applications in the personalized treatment of influenza.

Notation. $\Re$ is the set of real numbers, $\Re^{n\times m}$ is the set of $n \times m$ real matrices, and $\Re^n$ is the set of $n \times 1$ column vectors; $\aleph = \mathbb{Z}^+ \cup \{0\}$ is the set of nonnegative integers and $S \subset \aleph$ is the resetting set for control instants action. The set of $n \times n$ positive definite matrices is P^n, the matrix transpose is $(\cdot)^T$, the diagonal matrix is given by $diag(\cdot)$, and the inverse matrix is $(\cdot)^{-1}$.

6.5 Inverse optimal impulsive control

The impulsively controlled system has the form

$$\dot{x}(t) = f(x(t)), \; x_0 = x(0), \; t \neq \tau_j, \tag{6.1}$$

$$\Delta x(t) = f(x(\tau_j)) + g(x(\tau_j))\,\mathrm{sat}(u(\tau_j)), \; t = \tau_j, \tag{6.2}$$

$$\mathrm{sat}(u(\tau_j)) = \min\big(u(\tau_j), u_{max}(\tau_j)\big), \tag{6.3}$$

where $\tau_j \in S$ are the impulsive control instants, $x(\tau_j) \in \Re^n$ is the state at τ_j, and $x(t) \in \Re^n$ is the state of the system in $t \neq \tau_j$. The input $u(\tau_j) \in \Re^m$ and functions $f: \Re^n \to \Re^n$, $g: \Re^n \to \Re^{n\times m}$; $\Delta x(t)$ stands for the impulsive state modification at the control instant τ_j. The impulsive changes modify the initial condition for the following integration step with an instantaneous jump as follows:

$$x(\tau_j^+) = \Delta x(t)\big|_{t=\tau_j} + x(\tau_j^-), \tag{6.4}$$

$$\dot{x}(t) = f(x(\tau_j^+)), \quad t = \tau_j, \tag{6.5}$$

where $x(\tau_j^+)$ is the newly generated initial condition due to the impulsive change and $x(\tau_j^-)$ is the state before the impulsive jump. For the state modification with the impulsive law in (6.2), from the optimal control theory

[79,80], the cost functional

$$Q(x(\tau_j)) = \sum_{\tau_j \in S}^{\infty} \left(l(x(\tau_j)) + u(\tau_j)^T Ru(\tau_j) \right) \tag{6.6}$$

is minimized, where $Q : \Re^n \to \Re^+$, $l : \Re^n \to \Re^+$ weights the state vector $x(\tau_j)$ and is positive semidefinite. Besides, $R : \Re^n \to P^m$ weights the control effort.

Consider the impulsively controlled system (6.2) and, for simpler notation, $x(\tau_j^+)$ standing for the state change in the impulsive instant, the optimal control law [55,80]

$$u(\tau_j)^* = -\frac{1}{2} R^{-1} g(x(\tau_j))^T \frac{\partial Q^*(x(\tau_j^+))}{\partial(x(\tau_j^+))}, \tag{6.7}$$

which is inverse optimal if it satisfies the following: (1) it achieves exponential stability of the equilibrium point $x(\tau_j) = 0$ for system (6.2) (globally) with the optimal value function $Q^*(x(0)) = Q(x(0))$ and (2) it satisfies the cost functional (6.6), accounting for $l(x(\tau_j)) := -\overline{Q}$ and

$$\overline{Q} := u(\tau_j)^{*T} Ru(\tau_j)^* + Q(x(\tau_j^+)) - Q(x(\tau_j)) \leqslant 0. \tag{6.8}$$

Next, it follows that a function $Q(x(\tau_j))$ can be designed to satisfy (6.7)–(6.8). In this direction, consider $Q(x(\tau_j))$ in the following form [55]:

$$Q(x(\tau_j)) = \frac{1}{2} x(\tau_j)^T P_j x(\tau_j), \; P_j = P_j^T > 0, \tag{6.9}$$

which is used in the optimal control law (6.7) selecting a proper matrix $P_j \in P^n$. The inverse optimal impulsive control law is obtained applying (6.9) to (6.7) and optimizes the cost functional (6.6). Control law (6.7) is then modified to integrate the inverse approach as follows:

$$\begin{aligned}
u(\tau_j)^* &= -\frac{1}{2} R^{-1} g(x(\tau_j))^T \frac{\partial Q^*(x(\tau_j^+))}{\partial(x(\tau_j^+))} \\
&= -\frac{1}{2} R^{-1} g(x(\tau_j))^T \frac{\partial \left(\frac{1}{2} x(\tau_j^+)^T P_j x(\tau_j^+) \right)}{\partial(x(\tau_j^+))} \\
&= -\frac{1}{2} R^{-1} g(x(\tau_j))^T P_j \left(x(\tau_j) + f(x(\tau_j)) \right) \\
&\quad - \frac{1}{2} R^{-1} g(x(\tau_j))^T P_j g(x(\tau_j)) u(\tau_j)^*.
\end{aligned} \tag{6.10}$$

From (6.10), solving for $u(\tau_j)^*$,

$$u(\tau_j)^* + \frac{1}{2}R^{-1}g(x(\tau_j))^T P_j g(x(\tau_j))u(\tau_j)^* = -\frac{1}{2}R^{-1}g(x(\tau_j))^T P_j\big(x(\tau_j) + f(x(\tau_j))\big), \tag{6.11}$$

and multiplying (6.11) by R, it follows that

$$\big(R + \frac{1}{2}g(x(\tau_j))^T P_j g(x(\tau_j))\big)u(\tau_j)^* = -\frac{1}{2}g(x(\tau_j))^T P_j\big(x(\tau_j) + f(x(\tau_j))\big). \tag{6.12}$$

Finally, from (6.12), the inverse optimal impulsive control law $u(\tau_j)^*$ can be expressed as

$$u(\tau_j) = -\frac{1}{2}\big(R + P_\alpha(x(\tau_j))\big)^{-1} P_\beta(x(\tau_j)), \tag{6.13}$$
$$P_\alpha(x(\tau_j)) = \frac{1}{2}g(x(\tau_j))^T P_j g(x(\tau_j)),$$
$$P_\beta(x(\tau_j)) = g(x(\tau_j))^T P_j\big(x(\tau_j) + f(x(\tau_j))\big).$$

Importantly, the inverse matrix form in (6.13) is satisfied since $P_\alpha(x(\tau_j)$ is a symmetric and positive definite matrix [55,80]. The constraining input, sat($u(\tau_j)$), considers the maximum value $u_{max}(\tau_j)$ in (6.3); this is incorporated in terms of R in (6.13) as follows:

$$2u(\tau_j) = -\big(R + P_\alpha(x(\tau_j))\big)^{-1} P_\beta(x(\tau_j)),$$
$$2\big(R + P_\alpha(x(\tau_j))\big)u(\tau_j) = -P_\beta(x(\tau_j)),$$
$$2Ru(\tau_j) = -P_\beta(x(\tau_j)) - 2P_\alpha(x(\tau_j))u(\tau_j). \tag{6.14}$$

For a given u_{max}, the value of R in (6.14) is given by

$$R = -\frac{1}{2}\left[\frac{P_\beta(x(\tau_j)) + 2P_\alpha(x(\tau_j))u_{max}}{u_{max}}\right]. \tag{6.15}$$

Usually, the value of u_{max} is a priori given as, for instance, the maximum quantity of effort that the controller can handle due to physical or design limitations [50,78].

6.6 Influenza treatment tailoring

The control-based framework allows conceiving a control law whose structure incorporates the viral dynamics (or any other affine system) and

PK/PD effects. As shown in Fig. 6.5, whenever $t \neq \tau_j$ (solid arrow), the viral and PK/PD dynamics behave as an open loop; however, when $t = \tau_j$ (dotted arrows), at least one of the states of the PK/PD system is impulsively modified following (6.4), where the control effort is given by (6.13). In addition, using dynamic values of R in (6.15), the value of u_{max} considers dose constraints as, for instance, the maximum quantity of dose for a single intake. Finally, the drug intake frequency is no other than the impulsive instants; in other words, τ_j stands for the drug intake time (usually in hours).

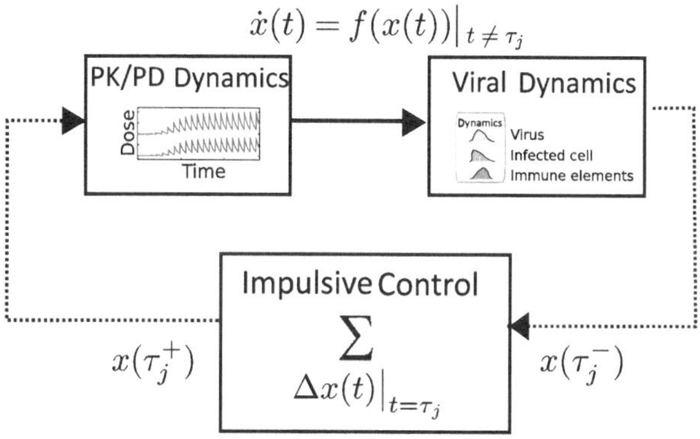

FIGURE 6.5 The control-based impulsive control. Whenever $t \neq \tau_j$, the viral and PK/PD dynamics behave as at open loop (solid arrow). The impulsive control modifies the PK/PD state of the system at $t = \tau_j$, which changes from $x(\tau_j^-)$ to $x(\tau_j^+)$ (dotted arrows), which corresponds with the drug dose.

Importantly, the PK/PD dynamics module in Fig. 6.5 is integrated by the PK and PD phases, where the PK phase is commonly constructed by at least one differential equation (compartment) describing the amount of available drug and its elimination rate. However, the PK phase can contain as many compartments as necessary to describe the drug distribution in organs and drug decomposition. The PD phase has a direct effect on the viral dynamics by means of the drug efficacy which is usually given by the EC_{50} value, which is the drug concentration level which provides 50% of drug efficacy. On the other hand, the viral dynamics module integrates diverse host dynamics that are important for the study of the infection and the treatment. In this framework, the PK/PD-based control computes the quantity of drug accounting for the drug compartments, drug efficacy, drug intake frequency, and host-related dynamics.

6.6.1 Numerical simulation of influenza tailoring treatment

The antiviral action dynamics represent the drug's impulsive change at every treatment intake (schedule), where the change magnitude is given by the control policy. The within-host model for influenza treatment considers the virus (V), the infected cells (I), and uninfected cells (U). The model writes as follows [77]:

$$\dot{V}(t) = p\left(1 - \frac{OC(t)}{OC(t) + EC_{50o}}\right) I(t) - cV(t), \tag{6.16}$$

$$\dot{U}(t) = -\beta U(t)V(t), \tag{6.17}$$

$$\dot{I}(t) = \beta U(t)V(t) - \delta I(t), \tag{6.18}$$

$$\dot{OP}(t) = k_a G(t) - k_f OP(t), \tag{6.19}$$

$$\dot{OC}(t) = k_f OP(t) - k_e OC(t), \tag{6.20}$$

$$\dot{G}(t) = -k_a G(t),\ t \neq \tau_j,\ \tau_j \leqslant t < \tau_{j+1}, \tag{6.21}$$

$$G(\tau_j^+) = U_G(\tau_j) + G(\tau_j^-),\ t = \tau_j. \tag{6.22}$$

Parameter values for Eqs. (6.16)–(6.21) are β (0.19×10^{-2} ml/d·TCID$_{50}$), δ (2.6 days^{-1}), p (70×10^{-5} TCID$_{50}$/d·ml), and c (6 days^{-1}). The initial conditions for the influenza system were taken from [81]; these are 1×10^{-5} TCID$_{50}$/ml for the virus, the number of susceptible cells is 4×10^8 cells, and zero for infected cells. The PK oseltamivir model is given by a three-compartment model representing the oseltamivir phosphate (OP), the active metabolic compound form oseltamivir carboxylate (OC), and the depot compartment (G) of dose administered before the adsorption in the blood at rate k_a [82]. The initial values of the drug's depot compartment, as well as $OP(0)$ and $OC(0)$, are zero (mg), changing only when the dose takes place. The parameter values of the drug are k_a (24.24 days^{-1}), k_f (16.41 days^{-1}), and k_e (3.26 days^{-1}); k_a is considered to be reduced with a decay rate of 7.31 after a dose intake [83,84]. On the other hand, the value of the oseltamivir PD phase, EC_{50o}, ranges between 0.0008 and >35 μM, where 1 μM = 0.284 μg/mL; 0.028 μM is considered, comparable to ca. 0.4 mg [42,77].

Control-based treatment and the current FDA recommended treatment with doses of 75 mg of oseltamivir are compared at τ_j, with $j = 12 \times 1, 2, \cdots$, standing for a drug intake each 12 h for 7 days. The control-based treatment considers $u_{max} = 75$ mg and the same impulsive instants τ_j. For the control law (6.13), the matrix $P_{j(IAV)} = diag(1)$, $P_{j(IAV)}(5,6) = P_{j(IAV)}(6,5) = 0.00039$. In addition, the order of the state vector is as follows: $x(t) = [G(t), CO(t), OP(t), U(t), I(t), V(t)]$ and $g(x(\tau_j)) = [0, 0, 0, 0, 0, 1]$ to form matrices $P_\alpha(x(\tau_j))$ and $P_\beta(x(\tau_j))$ of control law (6.13). The absolute value is applied to the control law and matrix R for positive values of drug amounts. Influenza treatment following FDA and control-based schemes manage to eliminate the viral load at 3 days postinfection (dpi) initiating treatment one dpi, shown in Fig. 6.6A. Importantly, while both schemes equally eliminate the viral load, the PK/PD

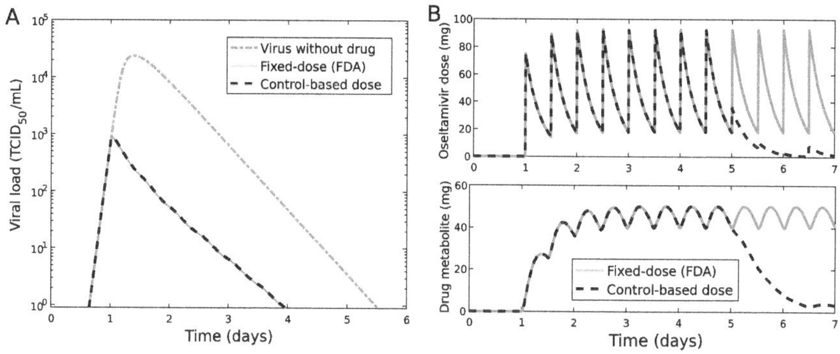

FIGURE 6.6 Influenza treatment tailoring. (A) Viral load, with the current FDA treatment scheme with fixed dose shown as a solid line, control-based treatment shown as a dashed line, and viral load without treatment shown as a dash-dotted line. (B) Drug dynamics, dynamics of the oseltamivir depot compartment G (top), and dynamics of the oseltamivir metabolite compartment (bottom). The control-based dose is reduced as the viral load in (A) goes down.

control-based treatment administered 30% less total drug (630 mg) than the FDA scheme (900 mg). Note that PK/PD-based treatment in Fig. 6.6B greatly decreases the drug input amounts after 5 dpi. Finally, the treatment efficacy is evaluated through the efficacy indicator E_f in (6.23) for both fixed-dose and control-based treatments, where they both reach more than 98% of efficacy. E_f is defined as follows:

$$E_f = 100\left(1 - \frac{VC_t}{VC}\right)\%, \tag{6.23}$$

where VC is the area under the viral curve without treatment and VC_t is the area under the viral curve with treatment.

6.7 Conclusions and final remarks

The path towards personalized treatment in influenza is still to be paved; however, pivotal points are explored in this chapter, covering from the global impact of pandemics in the last years to key aspects of the structure of the influenza virus. The pathogenesis, as well as a general view of the immune response to the infection, is also introduced. Regarding influenza treatment, the chapter provides a theoretical viewpoint towards treatment tailoring, accounting for a diversity of disciplines and combining inverse optimal, and impulsive control schemes that combine classical and novel theoretical bases concerning control-based stability, optimality, and impulsive action can be applied in the form of closed-loop system control. Results highlight that control-based treatment reaches the efficacy of fixed-dose medication; however, the drug quantity is considerably reduced, having important implications for treatment optimization and drug availability affairs, especially in pandemic settings.

In this sense, the significance of the proposed schemes can be extended to other acute infectious diseases such as COVID-19. First, testing treatment optimization of already existing drugs for other diseases as prospects for COVID-19 treatment [85], such as antivirals like Plitidepsin [86], Remdesivir [87], Baloxavir Marboxil, and Favipiravir [88], as well as antiinflammatory treatments [89,90], and the combination of them for diverse infection levels. Finally, in times of pandemic, as already experiencing with COVID-19 vaccine access, antiviral treatment optimization can pave the way of drug availability and accessibility to greater populations through individual treatment tailoring.

References

[1] WHO, World Health Organization. Influenza (Seasonal) fact sheet 2018, https://www.who.int/news-room/fact-sheets/detail/influenza-(seasonal), 2018. (Accessed February 2020).

[2] D. Molineux, Dr. Molineux's Historical account of the late general coughs and colds; with some observations on other Epidemick Distempers, Philosophical Transactions (1683-1775) (1694) 105–111.

[3] C.W. Potter, A history of influenza, Journal of Applied Microbiology 91 (2001) 572–579.

[4] R.C. Dicker, F. Coronado, D. Koo, R.G. Parrish, Principles of epidemiology in public health practice; an introduction to applied epidemiology and biostatistics, 2006.

[5] J.K. Taubenberger, J.C. Kash, D.M. Morens, The 1918 influenza pandemic: 100 years of questions answered and unanswered, Science Translational Medicine 11 (2019).

[6] P. Palese, T.M. Tumpey, A. Garcia-Sastre, What can we learn from reconstructing the extinct 1918 pandemic influenza virus?, Immunity 24 (2006) 121–124.

[7] F. Krammer, G.J. Smith, R.A. Fouchier, M. Peiris, K. Kedzierska, P.C. Doherty, P. Palese, M.L. Shaw, J. Treanor, R.G. Webster, et al., Influenza (primer), Nature Reviews Disease Primers (2018).

[8] D. Vijaykrishna, R. Mukerji, G.J. Smith, Rna virus reassortment: an evolutionary mechanism for host jumps and immune evasion, PLoS Pathogens 11 (2015) e1004902.

[9] M.C. White, A.C. Lowen, Implications of segment mismatch for influenza a virus evolution, Journal of General Virology 99 (2018) 3.

[10] F. Krammer, The human antibody response to influenza A virus infection and vaccination, Nature Reviews. Immunology (2019) 1.

[11] I. Kosik, J.W. Yewdell, Influenza hemagglutinin and neuraminidase: Yin–Yang proteins coevolving to thwart immunity, Viruses 11 (2019) 346.

[12] C. Ke, C.K.P. Mok, W. Zhu, H. Zhou, J. He, W. Guan, J. Wu, W. Song, D. Wang, J. Liu, et al., Human infection with highly pathogenic avian influenza A (H7N9) virus, China, Emerging Infectious Diseases 23 (2017) 1332.

[13] L.E. Leigh Perkins, D.E. Swayne, Pathogenicity of a Hong Kong–origin H5N1 highly pathogenic avian influenza virus for emus, geese, ducks, and pigeons, Avian Diseases 46 (2002) 53–63.

[14] R.W. Compans, M.B. Oldstone, Influenza Pathogenesis and Control-Volume I, Springer, 2014.

[15] T.H. Oguin, S. Sharma, A.D. Stuart, S. Duan, S.A. Scott, C.K. Jones, J.S. Daniels, C.W. Lindsley, P.G. Thomas, H.A. Brown, Phospholipase D facilitates efficient entry of influenza virus, allowing escape from innate immune inhibition, Journal of Biological Chemistry 289 (2014) 25405–25417.

[16] D.O. White, I.M. Cheyne, Early events in the eclipse phase of influenza and parainfluenza virus infection, Virology 29 (1966) 49–59.

[17] A. Boianelli, V.K. Nguyen, T. Ebensen, K. Schulze, E. Wilk, N. Sharma, S. Stegemann-Koniszewski, D. Bruder, F.R. Toapanta, C.A. Guzmán, et al., Modeling influenza virus infection: a roadmap for influenza research, Viruses 7 (2015) 5274–5304.

[18] J. Kreijtz, R. Fouchier, G. Rimmelzwaan, Immune responses to influenza virus infection, Virus Research 162 (2011) 19–30.

[19] S.-i. Tamura, T. Kurata, Defense mechanisms against influenza virus infection in the respiratory tract mucosa, Japanese Journal of Infectious Diseases 57 (2004) 236–247.

[20] I.K. Pang, A. Iwasaki, Inflammasomes as mediators of immunity against influenza virus, Trends in Immunology 32 (2011) 34–41.

[21] F. Heil, H. Hemmi, H. Hochrein, F. Ampenberger, C. Kirschning, S. Akira, G. Lipford, H. Wagner, S. Bauer, Species-specific recognition of single-stranded rna via toll-like receptor 7 and 8, Science 303 (2004) 1526–1529.

[22] H.M. Kim, Y.-W. Lee, K.-J. Lee, H.S. Kim, S.W. Cho, N. Van Rooijen, Y. Guan, S.H. Seo, Alveolar macrophages are indispensable for controlling influenza viruses in lungs of pigs, Journal of Virology 82 (2008) 4265–4274.

[23] C.H. GeurtsvanKessel, I.M. Bergen, F. Muskens, L. Boon, H.C. Hoogsteden, A.D. Osterhaus, G.F. Rimmelzwaan, B.N. Lambrecht, Both conventional and interferon killer dendritic cells have antigen-presenting capacity during influenza virus infection, PLoS ONE 4 (2009) e7187.

[24] J. Surls, C. Nazarov-Stoica, M. Kehl, S. Casares, T.-D. Brumeanu, Differential effect of cd4+ foxp3+ T-regulatory cells on the B and T helper cell responses to influenza virus vaccination, Vaccine 28 (2010) 7319–7330.

[25] A.G. Schmidt, M.D. Therkelsen, S. Stewart, T.B. Kepler, H.-X. Liao, M.A. Moody, B.F. Haynes, S.C. Harrison, Viral receptor-binding site antibodies with diverse germline origins, Cell 161 (2015) 1026–1034.

[26] J. Wrammert, K. Smith, J. Miller, W.A. Langley, K. Kokko, C. Larsen, N.-Y. Zheng, I. Mays, L. Garman, C. Helms, et al., Rapid cloning of high-affinity human monoclonal antibodies against influenza virus, Nature 453 (2008) 667–671.

[27] D. Angeletti, J.S. Gibbs, M. Angel, I. Kosik, H.D. Hickman, G.M. Frank, S.R. Das, A.K. Wheatley, M. Prabhakaran, D.J. Leggat, et al., Defining B cell immunodominance to viruses, Nature Immunology 18 (2017) 456.

[28] R. Nachbagauer, A. Choi, A. Hirsh, I. Margine, S. Iida, A. Barrera, M. Ferres, R.A. Albrecht, A. García-Sastre, N.M. Bouvier, et al., Defining the antibody cross-reactome directed against the influenza virus surface glycoproteins, Nature Immunology 18 (2017) 464.

[29] J. Sui, W.C. Hwang, S. Perez, G. Wei, D. Aird, L.-m. Chen, E. Santelli, B. Stec, G. Cadwell, M. Ali, et al., Structural and functional bases for broad-spectrum neutralization of avian and human influenza A viruses, Nature Structural & Molecular Biology 16 (2009) 265.

[30] G. Hernandez-Mejia, E.A. Hernandez-Vargas, Uncovering antibody cross-reaction dynamics in influenza A infections, Bioinformatics btaa691 (2020).

[31] S. Duwe, Influenza viruses–antiviral therapy and resistance, GMS Infectious Diseases 5 (2017).

[32] J.L. McKimm-Breschkin, Influenza neuraminidase inhibitors: antiviral action and mechanisms of resistance, Influenza and Other Respiratory Viruses 7 (2013) 25–36.

[33] G.N. Okoli, H.E. Otete, C.R. Beck, J.S. Nguyen-Van-Tam, Use of neuraminidase inhibitors for rapid containment of influenza: a systematic review and meta-analysis of individual and household transmission studies, PLoS ONE 9 (2014).

[34] J. Dobson, R.J. Whitley, S. Pocock, A.S. Monto, Oseltamivir treatment for influenza in adults: a meta-analysis of randomised controlled trials, The Lancet 385 (2015) 1729–1737.
[35] K. McClellan, C.M. Perry, Oseltamivir, Drugs 61 (2001) 263–283.
[36] A.C. Hurt, Antiviral therapy for the next influenza pandemic, Tropical Medicine and Infectious Disease 4 (2019) 67.
[37] M. Okomo-Adhiambo, A.M. Fry, S. Su, H.T. Nguyen, A.A. Elal, E. Negron, J. Hand, R.J. Garten, J. Barnes, X. Xiyan, et al., Oseltamivir-resistant influenza A (H1N1) pdm09 viruses, United States, 2013–14, Emerging Infectious Diseases 21 (2015) 136.
[38] F.G. Hayden, N. Sugaya, N. Hirotsu, N. Lee, M.D. de Jong, A.C. Hurt, T. Ishida, H. Sekino, K. Yamada, S. Portsmouth, et al., Baloxavir marboxil for uncomplicated influenza in adults and adolescents, The New England Journal of Medicine 379 (2018) 913–923.
[39] Y. Furuta, T. Komeno, T. Nakamura, Favipiravir (t-705), a broad spectrum inhibitor of viral RNA polymerase, Proceedings of the Japan Academy. Series B 93 (2017) 449–463.
[40] F.G. Hayden, N. Shindo, Influenza virus polymerase inhibitors in clinical development, Current Opinion in Infectious Diseases 32 (2019) 176.
[41] E. Takashita, C. Kawakami, R. Ogawa, H. Morita, S. Fujisaki, M. Shirakura, H. Miura, K. Nakamura, N. Kishida, T. Kuwahara, et al., Influenza A (H3N2) virus exhibiting reduced susceptibility to baloxavir due to a polymerase acidic subunit I38T substitution detected from a hospitalised child without prior baloxavir treatment, Japan, January 2019, Euro Surveillance 24 (2019).
[42] B. La Roche Ltd, Switzerland, Tamiflu (oseltamivir phosphate) capsules and oral suspension, https://www.accessdata.fda.gov/drugsatfda-docs/nda/2000/21-246-Tamiflu-Prntlbl.pdf, 2019, Online.
[43] P.H. Van der Graaf, N. Benson, Systems pharmacology: bridging systems biology and pharmacokinetics-pharmacodynamics (PK-PD) in drug discovery and development, Pharmaceutical Research 28 (2011) 1460–1464.
[44] T. Tängdén, V.R. Martín, T. Felton, E.I. Nielsen, S. Marchand, R.J. Brüggemann, J. Bulitta, M. Bassetti, U. Theuretzbacher, B. Tsuji, et al., The role of infection models and pk/pd modelling for optimising care of critically ill patients with severe infections, Intensive Care Medicine 43 (2017) 1021–1032.
[45] Y. Yu, D. Rüppel, W. Weber, H. Derendorf, PK/PD approaches, Drug Discovery and Evaluation: Methods in Clinical Pharmacology (2019) 1–23.
[46] I. Tavassoly, J. Goldfarb, R. Iyengar, Systems biology primer: the basic methods and approaches, Essays in Biochemistry 62 (2018) 487–500.
[47] S. Lucia, M. Schliemann-Bullinger, R. Findeisen, E. Bullinger, A set-based optimal control approach for pharmacokinetic/pharmacodynamic drug dosage design, IFAC-PapersOnLine 49 (2016) 797–802.
[48] G. Hernandez-Mejia, A.Y. Alanis, E.A. Hernandez-Vargas, Inverse optimal impulsive control based treatment of influenza infection, IFAC-PapersOnLine 50 (2017) 12185–12190.
[49] E.A. Hernandez-Vargas, Modeling and Control of Infectious Diseases in the Host: With MATLAB and R, Academic Press, 2019.
[50] G. Hernandez-Mejia, E.A. Hernandez-Vargas, PK/PD-based impulsive control to tailor therapies in infectious diseases, IFAC-PapersOnLine 53 (2020) 16055–16060.
[51] G. Montaseri, A. Boianelli, E.A. Hernandez-Vargas, M. Meyer-Hermann, PK/PD-based adaptive tailoring of oseltamivir doses to treat within-host influenza viral infections, Progress in Biophysics and Molecular Biology 139 (2018) 31–42.
[52] G.D. Vega-Magdaleno, A.Y. Alanis, G. Hernandez-Mejia, E.A. Hernandez-Vargas, Impulsive MPC for influenza infection treatment at variable time, IFAC-PapersOnLine 51 (2018) 79–84.

[53] J. Ferreira, E.A. Hernandez-Vargas, R.H. Middleton, Computer simulation of structured treatment interruption for HIV infection, Computer Methods and Programs in Biomedicine 104 (2011) 50–61.
[54] P.S. Rivadeneira, C.H. Moog, G.-B. Stan, C. Brunet, F. Raffi, V. Ferré, V. Costanza, M.J. Mhawej, F. Biafore, D.A. Ouattara, et al., Mathematical modeling of HIV dynamics after antiretroviral therapy initiation: a review, BioResearch Open Access 3 (2014) 233–241.
[55] E.N. Sanchez, F. Ornelas-Tellez, Discrete-Time Inverse Optimal Control for Nonlinear Systems, CRC Press, 2017.
[56] R. Sepulchre, M. Jankovic, P.V. Kokotovic, Constructive Nonlinear Control, Springer Science & Business Media, 2012.
[57] P. Werbos, Approximate dynamic programming for realtime control and neural modelling, in: Handbook of Intelligent Control: Neural, Fuzzy and Adaptive Approaches, 1992, pp. 493–525.
[58] F.-Y. Wang, H. Zhang, D. Liu, Adaptive dynamic programming: an introduction, IEEE Computational Intelligence Magazine 4 (2009) 39–47.
[59] A. Al-Tamimi, F.L. Lewis, M. Abu-Khalaf, Discrete-time nonlinear HJB solution using approximate dynamic programming: convergence proof, IEEE Transactions on Systems, Man and Cybernetics 38 (2008) 943–949.
[60] F.L. Lewis, D. Liu, Reinforcement Learning and Approximate Dynamic Programming for Feedback Control, Vol. 17, John Wiley & Sons, 2013.
[61] N. Ab Aza, A. Shahmansoorian, M. Davoudi, From inverse optimal control to inverse reinforcement learning: a historical review, Annual Reviews in Control (2020).
[62] B.D. Anderson, J.B. Moore, Optimal Control: Linear Quadratic Methods, Courier Corporation, 2007.
[63] R.A. Freeman, P.V. Kokotovic, Optimal nonlinear controllers for feedback linearizable systems, in: Proceedings of 1995 American Control Conference-ACC'95, Vol. 4, IEEE, 1995, pp. 2722–2726.
[64] R.E. Kalman, When is a linear control system optimal?, 1964.
[65] F. Ornelas, E.N. Sanchez, A.G. Loukianov, Discrete-time inverse optimal control for nonlinear systems trajectory tracking, in: 49th IEEE Conference on Decision and Control (CDC), IEEE, 2010, pp. 4813–4818.
[66] F. Ornelas-Tellez, E.N. Sanchez, A.G. Loukianov, Inverse optimal control for discrete-time nonlinear systems via passivation, Optimal Control Applications & Methods 35 (2014) 110–126.
[67] T.L. Molloy, J.J. Ford, T. Perez, Online inverse optimal control on infinite horizons, in: 2018 IEEE Conference on Decision and Control (CDC), IEEE, 2018, pp. 1663–1668.
[68] T.L. Molloy, J.J. Ford, T. Perez, Finite-horizon inverse optimal control for discrete-time nonlinear systems, Automatica 87 (2018) 442–446.
[69] H.K. Khalil, J.W. Grizzle, Nonlinear Systems, Vol. 3, Prentice Hall, Upper Saddle River, NJ, 2002.
[70] M. Krstic, H. Deng, A. Fettweis, J. Massey, M. Thoma, E. Sontag, B. Dickinson, Stabilization of Nonlinear Uncertain Systems, Springer-Verlag, 1998.
[71] J. Casti, On the general inverse problem of optimal control theory, Journal of Optimization Theory and Applications 32 (1980) 491–497.
[72] T. Yang, Impulsive Control Theory, Vol. 272, Springer Science & Business Media, 2001.
[73] T. Yang, Impulsive control, IEEE Transactions on Automatic Control 44 (1999) 1081–1083, https://doi.org/10.1109/9.763234.
[74] P.S. Rivadeneira, M. Caicedo, A. Ferramosca, A.H. González, Impulsive zone model predictive control (iZMPC) for therapeutic treatments: application to HIV dynamics, in: 2017 IEEE 56th Annual Conference on Decision and Control (CDC), IEEE, 2017, pp. 4094–4099.
[75] X. Li, D.W. Ho, J. Cao, Finite-time stability and settling-time estimation of nonlinear impulsive systems, Automatica 99 (2019) 361–368.

[76] X. Liu, P. Stechlinski, Infectious Disease Modeling, Springer, 2017.
[77] G. Hernandez-Mejia, A.Y. Alanis, M. Hernandez-Gonzalez, R. Findeisen, E.A. Hernandez-Vargas, Passivity-based inverse optimal impulsive control for influenza treatment in the host, IEEE Transactions on Control Systems Technology (2019).
[78] G. Hernandez-Mejia, X. Du, A.Y. Alanis, E.A. Hernandez-Vargas, Bounded input impulsive control for scheduling therapies, Journal of Process Control 102 (2021) 34–43.
[79] W.M. Haddad, V. Chellaboina, S.G. Nersesov, Impulsive and Hybrid Dynamical Systems: Stability, Dissipativity, and Control, Princeton University Press, 2006.
[80] G. Hernandez-Mejia, A.Y. Alanis, E.A. Hernandez-Vargas, Neural inverse optimal control for discrete-time impulsive systems, Neurocomputing 314 (2018) 101–108.
[81] A. Handel, I.M. Longini Jr, R. Antia, Neuraminidase inhibitor resistance in influenza: assessing the danger of its generation and spread, PLoS Computational Biology 3 (2007) e240.
[82] B.E. Davies, Pharmacokinetics of oseltamivir: an oral antiviral for the treatment and prophylaxis of influenza in diverse populations, Journal of Antimicrobial Chemotherapy 65 (2010) ii5–ii10.
[83] L. Canini, J.M. Conway, A.S. Perelson, F. Carrat, Impact of different oseltamivir regimens on treating influenza A virus infection and resistance emergence: insights from a modelling study, PLoS Computational Biology 10 (2014).
[84] Y. Wattanagoon, K. Stepniewska, N. Lindegårdh, S. Pukrittayakamee, U. Silachamroon, W. Piyaphanee, T. Singtoroj, W. Hanpithakpong, G. Davies, J. Tarning, et al., Pharmacokinetics of high-dose oseltamivir in healthy volunteers, Antimicrobial Agents and Chemotherapy 53 (2009) 945–952.
[85] A. Torneri, P. Libin, J. Vanderlocht, A.-M. Vandamme, J. Neyts, N. Hens, A prospect on the use of antiviral drugs to control local outbreaks of COVID-19, BMC Medicine 18 (2020) 1–9.
[86] K.M. White, R. Rosales, S. Yildiz, T. Kehrer, L. Miorin, E. Moreno, S. Jangra, M.B. Uccellini, R. Rathnasinghe, L. Coughlan, et al., Plitidepsin has potent preclinical efficacy against SARS-CoV-2 by targeting the host protein eEF1A, Science 371 (2021) 926–931.
[87] J.H. Beigel, K.M. Tomashek, L.E. Dodd, A.K. Mehta, B.S. Zingman, A.C. Kalil, E. Hohmann, H.Y. Chu, A. Luetkemeyer, S. Kline, D. Lopez de Castilla, R.W. Finberg, K. Dierberg, V. Tapson, et al., Remdesivir for the treatment of COVID-19 — final report, The New England Journal of Medicine 383 (2020) 1813–1826.
[88] Y. Lou, L. Liu, H. Yao, X. Hu, J. Su, K. Xu, R. Luo, X. Yang, L. He, X. Lu, et al., Clinical outcomes and plasma concentrations of baloxavir marboxil and favipiravir in COVID-19 patients: an exploratory randomized, controlled trial, European Journal of Pharmaceutical Sciences 157 (2021) 105631.
[89] P. Little, Non-steroidal anti-inflammatory drugs and COVID-19, 2020.
[90] W. Zhang, Y. Zhao, F. Zhang, Q. Wang, T. Li, Z. Liu, J. Wang, Y. Qin, X. Zhang, X. Yan, et al., The use of anti-inflammatory drugs in the treatment of people with severe coronavirus disease 2019 (COVID-19): the perspectives of clinical immunologists from China, Clinical Immunology 214 (2020) 108393.

CHAPTER

7

Polynomial state estimation in infectious diseases

M. Hernandez-Gonzalez

Autonomous University of Nuevo Leon, Department of Physical and Mathematical Sciences, San Nicolas de los Garza, Nuevo Leon, Mexico

7.1 Introduction

State estimation of linear and nonlinear stochastic systems has attracted the interest of many researchers over the last decades. One of those interests relies on control theory where many feedback controllers require the availability of the entire state. State estimation has found applications in several areas [3,14,15,25,27,29,34,30,40,43] and a wide variety of state estimation methods has been proposed such as the Wiener filter [36], the Kalman filter (KF) [23], the extended KF (EKF) [36], the adaptive filter [5], $\mathcal{H}_\infty$ Bayesian estimation [31], the particle filter [36], and the variable structure filter [2]. Among them, the most used algorithm for discrete-time linear stochastic systems in the presence of additive Gaussian noises and output linear measurements is the KF [23], which is a minimum-variance state estimator. This algorithm provides optimal state estimates in a recursive manner. A continuous-time version of the discrete-time KF can be found in the Kalman–Bucy filter [24]. Another well-known state estimator for linear systems is the $\mathcal{H}_\infty$ filter, which is a robust filter where the additive noise terms may have unknown statistics or they may be deterministic. In contrast to the KF, which requires an exact and accurate system model, the $\mathcal{H}_\infty$ filter requires no knowledge of noise statistics but assumes boundedness of the total noise energy. Another difference between the $\mathcal{H}_\infty$ filter and the KF algorithms is that $\mathcal{H}_\infty$ filter minimizes the influence of the worst possible disturbances on the estimation error and, consequently, is robust against model uncertainties [13]. In the case of nonlinear systems, the EKF is an alternative method due to its ease of implementation.

https://doi.org/10.1016/B978-0-32-390171-0.00016-0

It can be derived by linearizing the nonlinear system around its current state estimate [5] and then applying the KF equations. Other extensions of the KF to nonlinear systems are found in the unscented KF [2,13] and the cubature KF [2]. A transformation method to approximate the posterior Gaussian distribution by capturing its mean and covariance was proposed by [2]. In [17], a generalization of the KF to nonlinear systems has been proposed. The linearization steps are not involved and consequently, the derivations and evaluation of Jacobian matrices required by the EKF have been avoided. In [38], motivated by the large quantity of online operations required for nonlinear filtering problems, a Lyapunov-like method is employed to construct observers for continuous- and discrete-time nonlinear stochastic systems driven by noises with bounded covariances. Also, the conditions for the existence of an exponentially bounded observer for the nonlinear stochastic system are given. Basin et al. [8] presented the problem of state estimation for a class of nonlinear polynomial systems where the optimal filtering problem is treated proceeding from the general expression for the stochastic Itô differential of the optimal estimate and the error covariance. A finite set of equations can be obtained by expressing the conditional expectations of polynomial terms as functions of the estimate and estimation error covariance. A discrete-time state estimation for stochastic nonlinear polynomial systems driven by additive white Gaussian noises is presented in [19] where the estimation problem is separated in two parts: (1) computation of the time update and (2) computation of measurement update equations, for the state estimate and the error covariance matrix. In [1] a nonlinear KF is designed to estimate the states of a nonlinear large-scale power plant system. The nonlinear estimation approach is based on the state-dependent differential Riccati equation. The technique avoids linearization and can take into account the system nonlinearities and the measurement noise. The state estimation problem has applications in different disciplines such as in medicine [37], robotic applications [39], computer vision [42], speech enhancement [35], and electrical machines [11]. The paper [39] proposes a method for the safe autonomous positioning of an eye surgical robot combining a Gaussian mixture model for image segmentation and a KF to increase the accuracy. In [11] is reported a multiple EKF-based observer technique for the estimation of resistances R_r and R_s of induction motors. The approach is based on the use of two EKF algorithms derived from two extended models that results in the accurate estimation of a higher number of parameters that would be possible than with a single EKF algorithm. The paper [42] reports experimental results of a method for tracking in a sea search based on a KF and the CAMSHIFT algorithms.

On the other hand, high death tolls are derived from viral infections such as influenza, HIV, and Ebola. For instance, according to the World Health Organization, annual worldwide influenza epidemics are estimated to result in 3 to 5 million cases of severe illness and about 650,000 deaths [41]. A key obstacle to combat infectious diseases is the poor track of the pathogens and the respective immune responses in the host. Measurements in biology and biomedical systems are infrequent, expensive, difficult, and sometimes not possible to perform [4]. State estimation for biological nonlinear systems could play a central role in clinics providing online biological markers for a better understanding of infectious diseases. Surrogate marker estimations using nonlinear observers have received special attention in HIV infection [44,4]. However, nonlinear observers with application to infectious diseases remained largely unexplored. Thus, the development of estimation tools could bring the prognosis of infectious diseases and several other clinical applications to a new dimension. The observer design is a nonstandard problem in which past measurements should be propagated along unknown dynamics, which need to be corrected using a nonlinear output injection [7,12].

This chapter presents a solution for the state estimation problem of stochastic biological systems that are usually formulated as Michaelis–Menten and bilinear terms presented in different infections [22]. The main difference from previous works [9,10] is that herein the nonlinear polynomial function is represented in a more general form where matrices in the polynomial terms are state-dependent. Moreover, higher-order moments are obtained considering the relation between the characteristic function for an n-dimensional random vector and its moments. The proposed filter algorithm is compared with the results obtained by the EKF algorithm. Numerical simulations are performed in the influenza infection considering different mathematical models.

7.2 Preliminaries

Notation. The notation used throughout the chapter is standard: $\mathbb{R}^n$ denotes the n-dimensional Euclidean space, $\mathbb{R}^{n\times m}$ corresponds to the space of $n \times m$ matrices, and $\mathbb{R}^{n\times\cdots p \text{ times}\cdots n}$ is the space of p-dimensional tensors. For a given matrix (vector) A, $A^\top$ denotes its transpose.

Definitions

Stochastic process. [26] Suppose that with each element ω of a sample space Ω of some random experiment E, we associate a function $X(t,s)$, where t belongs to $T \subset \mathbb{R}$. The set $\{X(t,s), t \in T\}$ is called a stochastic (or random) process.

Wiener process. [33] A scalar or a vector random process with independent increments $W(t)$, $t > 0$, is called a Wiener process if it satisfies the following conditions:

a. all the realizations $w(t)$ of the process $W(t)$ are continuous and $w(0) = 0$;
b. the one-dimensional distribution of the process $W(t)$ is normal;

c. the expectation of the process $W(t)$ is equal identically to zero and its covariance function is given by the formula

$$K_w(t_1, t_2) = \int_0^{\min(t_1,t_2)} \upsilon(\tau)d\tau, \tag{7.1}$$

where $\upsilon(t)$ is a nonnegative function representing the intensity of the Wiener process $W(t)$.

Stochastic differential of the Itô process. [33] A random process $Z(t)$ is called the Itô process if it is representable in the form

$$Z(t) = Z(t_0) + \int_{t_0}^{t} X(\tau)d\tau + \int_{t_0}^{t} Y(\tau)dW(\tau), \tag{7.2}$$

where $t_0 > 0$ and $W(t)$ is the Wiener process. The first integral is the mean square integral and the second integral is the Itô stochastic integral; $X(t)$ and $Y(t)$ are random functions satisfying the conditions of the existence of the respective integrals.

The expression

$$dZ = X(t)dt + Y(t)dW(t) \tag{7.3}$$

is called the stochastic differential of the Itô process $Z(t)$.

Polynomial function. A polynomial of degree n is a function of the form

$$f(x) = a_0 + a_1x + a_2x^2 + \cdots + a_{n-1}x^{n-1} + a_nx^n, \tag{7.4}$$

where a_i, $i = 1, 2, \cdots, n$, are real numbers; a_i are also called the coefficients of the polynomial.

A generalization of the polynomial function in (7.4) can be expressed as follows:

$$f(x) = A_0 + A_1x + A_2xx + \cdots + A_pxx \cdots_{p\,\text{times}} \cdots x, \tag{7.5}$$

where $x \in \mathbb{R}^n$, $A_0 \in \mathbb{R}^n$, $A_1 \in \mathbb{R}^{n\times n}$, ..., $A_p \in \mathbb{R}^{n\times\cdots p+1\cdots\times n}$, and $x \times \cdots_{p\,\text{times}} \cdots \times x \in \mathbb{R}^{n\times\cdots p\cdots\times n}$. The rth element of the vector $f(x)$ is defined as

$$f_r(x(t)) = A_{0r} + \sum_i A_{1_{ri}}x_i + \sum_{ij} A_{2_{rij}}x_ix_j + \ldots + \sum_{i_1i_2\ldots i_p} A_{p_{ri_1i_2\ldots i_p}}x_{i_1}x_{i_2}\ldots x_{i_p}, \quad r, i, j, i_1, i_2, \ldots, i_p = 1, \ldots, n. \tag{7.6}$$

An example is given in Section 7.3.

7.3 Statement of the problem

Let $(\Omega, \mathcal{F}, P)$ be a complete probability space with an increasing right-continuous family of σ-algebras $\mathcal{F}_t$, $t \geq t_0$, and let $(W_1(t), \mathcal{F}_t, t \geq t_0)$ and $(W_2(t), \mathcal{F}_t, t \geq t_0)$ be independent Wiener processes. The $\mathcal{F}_t$-measurable random process $(x(t), y(t))$ is described by the following nonlinear differential equation with a polynomial term:

$$\begin{aligned} dx(t) &= f(x(t))dt + BdW_1(t), \\ x(t_0) &= x_0, \end{aligned} \tag{7.7}$$

and a linear differential equation for the observation process

$$dy(t) = (C_0 + Cx(t))dt + GdW_2(t), \tag{7.8}$$

where $x(t) \in \mathbb{R}^n$ is the state vector and $y(t) \in \mathbb{R}^m$ is the output linear observation. The initial condition $x_0 \in \mathbb{R}^n$ is a Gaussian random vector such that x_0, $W_1(t)$, and $W_2(t)$ are independent. It is assumed that $GG^\top$ is a positive definite matrix. Coefficients in (7.7) and (7.8) are constants of appropriate dimensions.

The nonlinear function $f(x(t))$ is assumed to present a polynomial form of n variables with state-dependent matrices and is introduced as follows [8]:

$$\begin{aligned} f(x(t)) = {} & A_0(x(t)) + A_1(x(t))x(t) + A_2(x(t))x(t)x(t) + \cdots \\ & + A_p(x(t))x(t)x(t) \cdots_{p\,\text{times}} \cdots x(t), \end{aligned} \tag{7.9}$$

where $A_0(x(t)) \in \mathbb{R}^n$, $A_1(x(t)) \in \mathbb{R}^{n\times n}$, ..., $A_p(x(t)) \in \mathbb{R}^{n\times \cdots p+1 \cdots \times n}$, and $x(t) \times \cdots_{p\ \text{times}} \cdots \times x(t) \in \mathbb{R}^{n\times \cdots p \cdots \times n}$. Matrices $A_i(x(t))$, $i = 0, \cdots, p$, are composed of nonpolynomial elements. The rth element of the vector $f(x(t))$ is defined as

$$\begin{aligned} f_r(x(t)) = {} & A_{0r}(x(t)) + \sum_i A_{1_{ri}}(x(t))x_i(t) + \\ & \sum_{ij} A_{2_{rij}}(x(t))x_i(t)x_j(t) + \ldots + \\ & \sum_{i_1 i_2 \ldots i_p} A_{p_{r i_1 i_2 \ldots i_p}}(x(t))x_{i_1}(t)x_{i_2}(t)\ldots x_{i_p}(t), \\ & r, i, j, i_1, i_2, \ldots, i_p = 1, \ldots, n. \end{aligned} \tag{7.10}$$

7.3.1 Example. Van der Pol equation

Consider the following third-degree nonlinear polynomial system:

$$dx_1(t) = x_2(t)dt + dW_1(t),$$

$$dx_2(t) = (-x_1(t) + \epsilon(1 - x_1^2(t))x_2(t))dt + dW_2(t), \tag{7.11}$$
$$x_1(t_0) = x_{10},$$
$$x_2(t_0) = x_{20},$$

where $x(t) = [x_1(t)\, x_2(t)]^\top \in \mathbb{R}^2$ is the state vector, x_{10} and x_{20} are the initial conditions, and $\epsilon \in \mathbb{R}$ is a constant parameter.

Eq. (7.11) can be expressed in the form given in (7.7), where $f(x(t))$ is given as

$$f(x(t)) = A_0 + A_1 x(t) + A_2 x(t)x(t) + A_3 x(t)x(t)x(t) + BdW(t).$$

According to Eq. (7.10),

$$A_0 = 0, \quad A_1 = \begin{bmatrix} 0 & 1 \\ -1 & \epsilon \end{bmatrix}, \quad A_2 = 0, \quad B = \begin{bmatrix} 1 & 0 \\ 0 & 1 \end{bmatrix}, \quad dW(t) = \begin{bmatrix} dW_1(t) \\ dW_2(t) \end{bmatrix},$$

$$A_3 = \begin{cases} -\epsilon, & \text{in position } r = 2, i_1 = 1, i_2 = 1, i_3 = 2, \\ 0, & \text{others,} \end{cases} \quad \text{i.e., } A_{3_{[2112]}} = -\epsilon,$$

and other positions are zero.

7.4 Solution of the problem

The estimation problem consists in finding an estimate $\hat{x}(t)$ of the system state $x(t)$, based on the observation process $Y(t) = \{y(s), 0 \le s \le t\}$, that minimizes the Euclidean 2-norm [8]:

$$J = \mathbb{E}[(x(t) - \hat{x}(t))^\top (x(t) - \hat{x}(t)) | \mathcal{F}_t^Y] \tag{7.12}$$

at every time moment t. Here, $\mathbb{E}[z(t)|\mathcal{F}_t^Y]$ means the conditional expectation of a stochastic process $z(t) = (x(t) - \hat{x}(t))^\top (x(t) - \hat{x}(t))$ with respect to the σ-algebra $\mathcal{F}_t^Y$ generated by the observation process $Y(t)$ in the interval $[t_0, t]$. This estimate is given by the conditional expectation

$$\hat{x}(t) = \mathbb{E}[x(t)|\mathcal{F}_t^Y] \tag{7.13}$$

of the system state $x(t)$ with respect to the σ-algebra F_t^Y generated by the observation process $Y(t)$ in the interval $[t_0, t]$. Eq. (7.13) is complemented with the estimation error covariance matrix

$$P(t) = \mathbb{E}[(x(t) - \hat{x}(t))(x(t) - \hat{x}(t))^\top | \mathcal{F}_t^Y]. \tag{7.14}$$

7.4.1 Filtering equations for polynomial systems over linear observations

The filtering equations are obtained using the formula for the Itô differential of the conditional expectation $\hat{x}(t) = \mathbb{E}[x(t)|F_t^Y]$ and the covariance function $P(t) = \mathbb{E}[(x(t) - \hat{x})(x(t) - \hat{x}(t))^\top|F_t^Y]$ (see [8] for details):

$$\begin{aligned} d\hat{x}(t) = {} & \mathbb{E}(f(x(t))|\mathcal{F}_t^Y)dt + \mathbb{E}(x(t)[\varphi_1(x(t)) - \\ & E(\varphi_1(x(t))|\mathcal{F}_t^Y)]^\top|\mathcal{F}_t^Y)(GG)^{-1} \\ & \times (dy(t) - E(\varphi_1(x(t))|\mathcal{F}_t^Y)dt), \end{aligned} \tag{7.15}$$

where $f(x(t))$ is the polynomial term in Eq. (7.7), $\varphi_1(x) = C_0 + Cx(t)$. Substituting $\varphi_1(x(t))$ in Eq. (7.15), we obtain

$$\begin{aligned} d\hat{x}(t) = {} & \mathbb{E}(f(x(t))|\mathcal{F}_t^Y)dt + \mathbb{E}(x(t)[C_0 + Cx(t) \\ & - E(C_0 + Cx(t)|\mathcal{F}_t^Y)]^\top|\mathcal{F}_t^Y) \\ & \times (GG)^{-1}(dy(t) - E(C_0 + Cx(t)|\mathcal{F}_t^Y)dt) \\ = {} & \mathbb{E}(f(x(t))|\mathcal{F}_t^Y)dt + \mathbb{E}(x(t)[Cx(t) \\ & - C\hat{x}(t)|\mathcal{F}_t^Y)]^\top|\mathcal{F}_t^Y)(GG)^{-1}(dy(t) - (C_0 + C\hat{x}(t))dt) \\ = {} & \mathbb{E}(f(x(t))|\mathcal{F}_t^Y)dt + \mathbb{E}(x(t)[x(t) \\ & - \hat{x}(t)|\mathcal{F}_t^Y)]^\top|\mathcal{F}_t^Y)C^\top(GG)^{-1}(dy(t) - (C_0 + C\hat{x}(t))dt) \\ = {} & \mathbb{E}(f(x(t))|\mathcal{F}_t^Y)dt + P(t)C^\top(GG^\top)^{-1}(dy(t) - (C_0 + C\hat{x}(t))dt). \end{aligned} \tag{7.16}$$

In Eq. (7.16) we have used the fact that

$$P(t) = \mathbb{E}[(x(t) - \hat{x})(x(t) - \hat{x}(t))^\top|F_t^Y]$$

and

$$\mathbb{E}[(x(t)x^\top(t)|F_t^Y] = P(t) + \hat{x}(t)\hat{x}^\top(t).$$

Eq. (7.16) is complemented with the initial condition

$$\hat{x}(t_0) = E[x(t_0)|F_{t_0}^Y].$$

Note that Eq. (7.16) should be complemented with the equation of the error covariance matrix $P(t)$. The equation for the covariance matrix $P(t)$ is given by (see [8,33])

$$\begin{aligned} dP(t) = {} & (\mathbb{E}((x(t) - \hat{x}(t))f^\top(x(t))|\mathcal{F}_t^Y) \\ & + \mathbb{E}(f(x(t))(x(t) - \hat{x}(t))^\top|\mathcal{F}_t^Y) + BB^\top \end{aligned} \tag{7.17}$$

$$- \mathbb{E}(x(t)[\varphi_1(x) - \mathbb{E}(\varphi_1(x)|\mathcal{F}_t^Y)]^\top|\mathcal{F}_t^Y)(GG^\top)^{-1}$$
$$\times \mathbb{E}([\varphi_1(x) - E(\varphi_1(x)|\mathcal{F}_t^Y)]x^\top(t)|\mathcal{F}_t^Y))dt.$$

Upon substituting the expression for $\varphi_1(x)$, Eq. (7.17) takes the form

$$\begin{aligned} dP(t) = {} & (\mathbb{E}((x(t) - \hat{x}(t))f^\top(x(t))|\mathcal{F}_t^Y) \\ & + \mathbb{E}(f(x(t))(x(t) - \hat{x}(t))^\top|\mathcal{F}_t^Y) + BB^\top \\ & - \mathbb{E}(x(t)[C_0 + Cx(t) - \mathbb{E}(C_0 + Cx(t)|\mathcal{F}_t^Y)]^\top|\mathcal{F}_t^Y)(GG^\top)^{-1} \\ & \times \mathbb{E}([C_0 + Cx(t) - \mathbb{E}(C_0 + Cx(t)|\mathcal{F}_t^Y)]x^\top(t)|\mathcal{F}_t^Y))dt \\ = {} & (\mathbb{E}((x(t) - \hat{x}(t))f^\top(x(t))|\mathcal{F}_t^Y) + \mathbb{E}(f(x(t))(x(t) - \hat{x}(t))^\top|\mathcal{F}_t^Y) \\ & + BB^\top - \mathbb{E}(x(t)[x(t) - \hat{x}(t)]^\top|\mathcal{F}_t^Y)C^\top(GG^\top)^{-1} \\ & \times C\mathbb{E}([x(t) - \hat{x}(t)]x^\top(t)|\mathcal{F}_t^Y))dt \\ = {} & (\mathbb{E}((x(t) - \hat{x}(t))f^\top(x(t))|\mathcal{F}_t^Y) + \mathbb{E}(f(x(t))(x(t) - \hat{x}(t))^\top|\mathcal{F}_t^Y) \\ & + BB^\top - P(t)C^\top(G(t)G^\top(t))^{-1}CP(t))dt, \end{aligned} \tag{7.18}$$

with the corresponding initial conditions

$$\begin{aligned} \hat{x}(t_0) &= \mathbb{E}(x(t_0)|F_{t_0}^Y), \\ P(t_0) &= \mathbb{E}[(x(t_0) - \hat{x}(t_0))(x(t_0) - \hat{x}(t_0))^\top|F_{t_0}^Y]. \end{aligned} \tag{7.19}$$

The filtering equations for the nonlinear system (7.7)–(7.9) are governed by Eqs. (7.16) and (7.18). The closed-form solution is obtained once the conditional expectations $\mathbb{E}[f(x(t))|F_t^Y]$, $\mathbb{E}[(x(t) - \hat{x}(t))f^\top(x(t))|\mathcal{F}_t^Y]$, and $\mathbb{E}(f(x(t))(x(t) - \hat{x}(t))^\top|\mathcal{F}_t^Y)$ are expressed in terms of the expectation $\hat{x}(t)$ and the covariance matrix $P(t)$. This follows the property of the conditionally Gaussian random variable $x(t) - \hat{x}(t)$: all its odd conditional moments $\mathbb{E}[x(t) - \hat{x}(t)|F_t^Y]$, $\mathbb{E}[(x(t) - \hat{x}(t))^3|F_t^Y]$, and $\mathbb{E}[(x(t) - \hat{x}(t))^5|F_t^Y]$, ... are equal to zero, and all its even conditional moments $\mathbb{E}[(x(t) - \hat{x}(t))^2|F_t^Y] = P(t)$, $\mathbb{E}[(x(t) - \hat{x}(t))^4|F_t^Y] = 3P^2$, ... are represented as functions of the covariance matrix $P(t)$ (see [8] for details). Appropriate approximations of expectations for nonpolynomial functions $A_i(x(t)) \approx A_i(\hat{x}(t))$, $i = 1, \cdots, p$, have been also taken into account.

Considering the previous comment and that $f(x(t))$ in Eq. (7.9) has polynomial terms, the conditional expectations for each element in $f(x(t))$, $(x(t) - \hat{x}(t))f^\top(x(t))$, $f(x(t))(x(t) - \hat{x}(t))^\top$ are derived for the rth-order moment of an n-dimensional normal random vector as follows [32]:

$$\alpha_{h_1,\cdots,h_n}(t) = j^{-r}\left[\frac{\partial^r}{\partial w_1^{h_1}\cdots\partial w_n^{h_n}}\Psi(w;t)\right]_{w=0}, \tag{7.20}$$

where $\Psi(w;t) = e^A$ and

$$A = jw^\top \hat{x}(t) - \frac{1}{2} w^\top P(t) w, \qquad j = \sqrt{-1}, \tag{7.21}$$

with

$$w = [w_1\, w_2 \,\cdots\, w_n]^\top, \qquad x(t) = [x_1(t)\, x_2(t) \,\cdots\, x_n(t)]^\top,$$
$$\hat{x}(t) = [\hat{x}_1(t)\, \hat{x}_2(t) \,\cdots\, \hat{x}_n(t)]^\top, \qquad h_1 + h_2 + \cdots + h_n = r.$$

7.4.1.1 *Example*

Consider the following nonlinear function with two variables:

$$f(x_1, x_2) = x_1^2 x_2, \tag{7.22}$$

where its expectation is computed according to (7.20). Note that the polynomial function in (7.22) corresponds to a term in the second equation in (7.11).

Let us define the function $\Psi(w;t) = e^A$, with $A = jw^\top \hat{x}(t) - \frac{1}{2} w^\top P(t) w$, $h_1 = 2$, $h_2 = 1$, $r = 3$, where

$$w = [w_1 \;\, w_2], \qquad \hat{x}(t) = [x_1(t) \;\, x_2(t)], \qquad P(t) = \begin{bmatrix} P_{11}(t) & P_{12}(t) \\ P_{12}(t) & P_{22}(t) \end{bmatrix}.$$

A in (7.21) takes the form

$$A = j(w_1 \hat{x}_1 + w_2 \hat{x}_2) - \frac{1}{2}(w_1^2 P_{11} + 2 w_1 w_2 P_{12} + w_2^2 P_{22}).$$

Upon performing the corresponding differentiations, the following expectation of

$$\mathbb{E}[f(x_1, x_2)] = \mathbb{E}[x_1^2 x_2] = \alpha_{2,1} = j^{-3} \left[\frac{\partial^3}{\partial w_1^2 \partial w_2} \Psi(w;t) \right]_{w=0}$$

is obtained:

$$\begin{aligned}
\frac{\partial \Psi(w;t)}{\partial w_2} &= e^A (j \hat{x}_2(t) - P_{22} w_2 - P_{12} w_1), \\
\frac{\partial^2 \Psi(w;t)}{\partial w_1 w_2} &= e^A (-\hat{x}_1(t)\hat{x}_2(t) - j P_{12}(t) w_2 \hat{x}_2(t) - j P_{11}(t) w_1 \hat{x}_2(t) \\
&\quad - j P_{22}(t) w_2 \hat{x}_1(t) - j P_{12}(t) w_1 \hat{x}_1(t) + P_{12}(t) P_{22}(t) w_2^2 \\
&\quad + P_{11}(t) P_{22}(t) w_1 w_2 + P_{12}^2(t) w_1 w_2 + P_{11}(t) P_{12} w_1^2 + P_{12}(t)),
\end{aligned}$$

$$
\begin{aligned}
\frac{\partial^3\Psi(w;t)}{\partial w_1^2 w_2} = e^A(&-j\hat{x}_1^2(t)\hat{x}_2(t) + 2P_{12}(t)w_2\hat{x}_1(t)\hat{x}_2(t) + 2P_{11}(t)w_1\hat{x}_1(t)\hat{x}_2(t)\\
&+ jP_{12}^2(t)w_2^2\hat{x}_2(t) + 2jP_{11}(t)P_{12}(t)w_1w_2\hat{x}_2(t) + jP_{11}^2(t)w_1^2\hat{x}_2(t)\\
&- jP_{11}(t)\hat{x}_2(t) + P_{22}w_2\hat{x}_1^2(t) + P_{12}(t)w_1\hat{x}_1^2(t)\\
&+ 2jP_{12}(t)P_{22}(t)w_2^2\hat{x}_1(t) + 2jP_{11}(t)P_{22}(t)w_1w_2\hat{x}_1(t)\\
&+ 2jP_{12}^2(t)w_1w_2\hat{x}_1(t) + 2jP_{11}(t)P_{12}(t)w_1^2\hat{x}_1(t) - 2jP_{12}(t)\hat{x}_1(t)\\
&- P_{12}^2(t)P_{22}(t)w_2^3 - 2P_{11}(t)P_{12}(t)P_{22}(t)w_1w_2^2 - P_{12}^3(t)w_1w_2^2\\
&- P_{11}^2(t)P_{22}w_1^2w_2 - 2P_{11}(t)P_{12}^2(t)w_1^2w_2 + P_{11}(t)P_{22}(t)w_2\\
&+ 2P_{12}^2(t)w_2 - P_{11}^2(t)P_{12}(t)w_1^3 + 3P_{11}(t)P_{12}(t)w_1).
\end{aligned}
$$

Finally,

$$
\begin{aligned}
\mathbb{E}[x_1^2x_2] &= j^{-3}\left[\frac{\partial^3}{\partial w_1^2\partial w_2}\Psi(w;t)\right]_{w=0}\\
&= j^{-3}(-j\hat{x}_1^2(t)\hat{x}_2(t) - jP_{11}(t)\hat{x}_2(t) - 2jP_{12}(t)\hat{x}_1(t))\\
&= \hat{x}_1^2(t)\hat{x}_2(t) + P_{11}(t)\hat{x}_2(t) + 2P_{12}(t)\hat{x}_1(t).
\end{aligned}
\tag{7.23}
$$

Remark. Once the nonlinear function $f(x(t))$ is expressed into a polynomial form with state-dependent matrices (7.9), an approximate estimation of the ith term is obtained as follows:

$$
\begin{aligned}
&\mathbb{E}[A_i(x(t))x(t)x(t)\ldots_{i\text{ times}}\ldots x(t)] \approx\\
&A_i(\hat{x}(t))\mathbb{E}(x(t)x(t)\ldots_{i\text{ times}}\ldots x(t)),\ i = 1,\cdots,p,
\end{aligned}
$$

i.e., the expectation of the state-dependent matrix $A_i(x(t))$ which has nonpolynomial terms is approximated replacing the vector state $x(t)$ by its estimation $\hat{x}(t)$. The remainder expectation $\mathbb{E}(x(t)x(t)\ldots_{i\text{ times}}\ldots x(t))$ is obtained from (7.20)–(7.21) taking into account the comments after Eq. (7.19).

7.4.2 Second-order polynomial function

Consider a particular case where $f(x(t))$ is defined as a second-degree polynomial function:

$$
f(x(t)) = A_0(x(t)) + A_1(x(t))x(t) + A_2(x(t))x(t)x(t),
$$

where $A_0(x(t))$, $A_1(x(t))$, and $A_3(x(t))$ are previously defined nonpolynomial functions. The finite-dimensional filtering equations (7.16) and (7.18) take the following form:

$$\begin{aligned} d\hat{x}(t) = (A_0(\hat{x}(t)) + A_1(\hat{x}(t))\hat{x}(t) + A_2(\hat{x}(t))\hat{x}(t)\hat{x}(t) + A_2(\hat{x}(t))P(t))dt \\ + P(t)C^\top(GG^\top)^{-1}(dy(t) - (C_0 + C\hat{x}(t))dt), \end{aligned} \tag{7.24}$$

$$\begin{aligned} dP(t) = (A_1(\hat{x}(t))P(t) + P(t)A_1^\top(\hat{x}(t)) + 2A_2(\hat{x}(t))\hat{x}(t)P(t) \\ + 2(A_2(\hat{x}(t))\hat{x}(t)P(t))^\top + BB^\top \\ - P(t)C^\top(GG^\top)^{-1}CP(t))dt, \end{aligned} \tag{7.25}$$

with the corresponding initial conditions

$$\begin{aligned} \hat{x}(t_0) = \mathbb{E}(x(t_0)|F_{t_0}^Y), \\ P(t_0) = \mathbb{E}[(x(t_0) - \hat{x}(t_0))(x(t_0) - \hat{x}(t_0))^\top|F_{t_0}^Y]. \end{aligned} \tag{7.26}$$

7.4.3 Example

Consider the following set of equations:

$$\begin{aligned} dx(t) = \sin(x(t))x^2(t)dt + dW_1(t), \\ dy(t) = x(t) + dW_2(t), \end{aligned} \tag{7.27}$$

with initial condition $x(0) = x_0$.

According to Eqs. (7.24) and (7.25) with $A_0(x(t)) = 0$, $A_1(x(t)) = 0$, $A_2(x(t)) = \sin(x(t))$, $G = 1$, $C_0 = 0$, $C = 1$, the following filtering equations are obtained:

$$\begin{aligned} d\hat{x}(t) = (\sin(\hat{x}(t))\hat{x}^2(t) + \sin(\hat{x}(t))P(t))dt + P(t)(dy(t) - \hat{x}(t)), \\ dP(t) = (4\sin(\hat{x}(t))\hat{x}(t)P(t) + 1 - P^2(t))dt, \end{aligned} \tag{7.28}$$

with initial conditions $\hat{x}(0) = \mathbb{E}[x_0|F_{t_0}^Y]$ and $P(0) = \mathbb{E}[(x_0 - \hat{x}(0))^2|F_{t_0}^Y]$.

The next section presents an application of the obtained filtering equations to a mathematical model for biological models.

7.5 State estimation for systems of viral infections

To test the performance of the proposed filter, numerical simulations are carried out employing two mathematical models for viral infections, the target cell and the virus immune response models.

7.5.1 Target cell model: general infectious disease system

The target cell model is set as a mathematical generality for viral infection diseases. It extends the original model from [6] in its stochastic form,

$$
\begin{aligned}
dV(t) &= (pI(t) - cV(t))dt + dW_V(t), \\
dU(t) &= -\beta U(t)V(t)dt + dW_U(t), \\
dI(t) &= (\beta U(t)V(t) - \delta I(t))dt + dW_I(t), \\
dy(t) &= V(t)dt + dW_y(t),
\end{aligned}
\tag{7.29}
$$

where U is the number of uninfected target cells, I is the number of productively infected cells, and V is the infectious viral titer; $W_V(t)$, $W_U(t)$, $W_I(t)$, and $W_y(t)$ are independent Wiener processes. Susceptible cells become infected by virus at a rate βUV, where β is the rate constant characterizing infection. Infected cells (I) replicate virus at an average rate p and die at a rate δ. Free virus is cleared at a rate c [6].

Parameter values are as follows: β (2.7×10^{-5} $(\mathrm{TCID_{50}/ml})^{-1}$ d^{-1}), p (1.2×10^{-2} $(\mathrm{TCID_{50}/ml})$ d^{-1}), c (3 d^{-1}), and δ (4 d^{-1}). Initial numbers of model variables are $V_0 = 1 \times 10^{-2}$ $\mathrm{TCID_{50}/ml}$, $U_0 = 3.95 \times 10^8$ cells, and $I_0 = 0$ cells [6]. Eq. (7.29) can be represented into a polynomial form (7.9) where

$$
f(x(t)) = A_1 x(t) + A_2 x(t) x(t),
$$

$$
A_1 = \begin{pmatrix} pI(t) - cV(t) \\ 0 \\ -\delta I(t) \end{pmatrix}, \quad A_2 x(t) x(t) = \begin{pmatrix} 0 \\ -\beta U(t)V(t) \\ \beta U(t)V(t) \end{pmatrix}.
\tag{7.30}
$$

Upon substituting Eq. (7.30) into Eqs. (7.16) and (7.18) and using (7.20) and (7.21) or (7.24) and (7.25), the polynomial state estimation for the target cell system (7.29) is given as follows.

7.5.1.1 Polynomial state estimation for the target cell system

We have obtained the following equations:

$$
\begin{aligned}
d\hat{V}(t) &= (-c\hat{V}(t) + p\hat{I}(t))dt + G^{-2}P_{11}(t)(dy(t) - \hat{V}(t)dt), \\
d\hat{U}(t) &= (-\beta\hat{U}(t)\hat{V}(t) - \beta P_{12}(t))dt + G^{-2}P_{12}(t)(dy(t) \\
&\quad - \hat{V}(t)dt), \\
d\hat{I}(t) &= (-\delta\hat{I}(t) + \beta\hat{U}(t)\hat{V}(t) + \beta P_{12}(t))dt \\
&\quad + G^{-2}P_{13}(t)(dy(t) - \hat{V}(t)dt),
\end{aligned}
\tag{7.31}
$$

where

$$
\begin{aligned}
dP_{11}(t) &= (-2cP_{11}(t) + 2pP_{13}(t) + b_1^2 - G^{-2}P_{11}^2(t))dt, \\
dP_{12}(t) &= (-cP_{12}(t) + pP_{23}(t) - \beta\hat{V}(t)P_{12}(t) \\
&\quad - \beta\hat{U}(t)P_{11}(t) - G^{-2}P_{11}(t)P_{12}(t))dt,
\end{aligned}
$$

$$
\begin{aligned}
dP_{13}(t) &= (-cP_{13}(t) - \delta P_{13} + pP_{33}(t) \\
&\quad + \beta\hat{V}(t)P_{12}(t) + \beta\hat{U}(t)P_{11}(t) \\
&\quad - G^{-2}P_{11}(t)P_{13}(t))dt, \\
dP_{22}(t) &= (-2\beta\hat{V}(t)P_{22}(t) - 2\beta\hat{U}(t)P_{12}(t) + b_2^2 \\
&\quad - G^{-2}P_{12}^2(t))dt, \\
dP_{23}(t) &= (-\delta P_{23}(t) - \beta\hat{V}(t)P_{23}(t) - \\
&\quad \beta\hat{U}(t)P_{13}(t) + \beta\hat{V}(t)P_{22}(t) + \beta\hat{U}(t)P_{12}(t) \\
&\quad - G^{-2}P_{12}(t)P_{13}(t))dt, \\
dP_{33}(t) &= (-2\delta P_{33}(t) + 2\beta\hat{V}(t)P_{23}(t) + 2\beta\hat{U}(t)P_{13}(t) \\
&\quad + b_3^2 - G^{-2}P_{13}^2(t))dt.
\end{aligned}
\tag{7.32}
$$

7.5.1.2 Extended Kalman filter equations

The EKF has been a very helpful state estimator for nonlinear systems. It takes the nonlinear system and linearizes it around the current state estimate. The filtering equations are then obtained by employing the KF equations on the linearized equations.

Consider the following nonlinear system:

$$
\begin{aligned}
dx(t) &= f(x(t))dt + BdW_1(t), \\
x(t_0) &= x_0
\end{aligned}
\tag{7.33}
$$

and a linear differential equation for the observation process,

$$
dy(t) = (C_0 + Cx(t))dt + GdW_2(t), \tag{7.34}
$$

where $x(t) \in \mathbb{R}^n$ is the state vector, $y(t) \in \mathbb{R}^m$ is the output linear observation, and $W_1(t)$ and $W_2(t)$ are independent Wiener processes. The initial condition $x_0 \in \mathbb{R}^n$ is a Gaussian random vector such that x_0, $W_1(t)$, and $W_2(t)$ are independent.

The EKF equations can be obtained computing the derivative matrices evaluated at the current state estimate

$$
A = \left.\frac{\partial f}{\partial x}\right|_{x(t)=\hat{x}(t)}, \qquad C = \frac{\partial(C_0 + Cx(t))}{\partial x}
$$

and executing the KF equations

$$
\begin{aligned}
d\hat{x}(t) &= f(\hat{x}(t))dt + PC^\top(GG^\top)^{-1}(dy(t) - (C_0 + C\hat{x}(t))dt), \\
dP(t) &= (AP(t) + P(t)A^\top + BB^\top - P(t)C^\top(GG^\top)^{-1}CP(t))dt
\end{aligned}
\tag{7.35}
$$

with the corresponding initial conditions

$$\begin{aligned}\hat{x}(t_0) &= \mathbb{E}(x(t_0)|F_{t_0}^Y),\\ P(t_0) &= \mathbb{E}[(x(t_0)-\hat{x}(t_0))(x(t_0)-\hat{x}(t_0))^\top|F_{t_0}^Y].\end{aligned} \tag{7.36}$$

For comparison purposes, the estimation equations for the target cell model 7.29 are obtained employing the EKF equations (7.35) and (7.36):

$$\begin{aligned}d\hat{V}_{ekf}(t) &= (p\hat{I}_{ekf}(t) - c\hat{V}_{ekf}(t))dt\\ &\quad + G^{-2}P_{11}(t)(dy(t) - \hat{V}_{ekf}(t)dt),\\ d\hat{U}_{ekf}(t) &= (-\beta\hat{U}_{ekf}(t)\hat{V}_{ekf}(t))dt\\ &\quad + G^{-2}P_{12}(t)(dy(t) - \hat{V}_{ekf}(t)dt),\\ d\hat{I}_{ekf}(t) &= (-\delta\hat{I}_{ekf}(t) + \beta\hat{U}_{ekf}(t)\hat{V}_{ekf}(t))dt\\ &\quad + G^{-2}P_{13}(t)(dy(t) - \hat{V}_{ekf}(t)dt),\end{aligned} \tag{7.37}$$

$$\begin{aligned}dP_{11ekf}(t) &= (2pP_{13ekf}(t) - 2cP_{11ekf}(t)\\ &\quad - G^{-2}P_{11ekf}^2(t) + b_1^2)dt,\\ dP_{12ekf}(t) &= (pP_{23ekf}(t) - cP_{12ekf}(t)\\ &\quad - \beta P_{11ekf}(t)\hat{U}_{ekf}(t) - \beta P_{12ekf}(t)\hat{V}_{ekf}(t)\\ &\quad - G^{-2}P_{11ekf}(t)P_{12ekf}(t))dt,\\ dP_{13ekf}(t) &= (pP_{33ekf}(t) - \delta P_{13ekf}(t) - cP_{13ekf}(t)\\ &\quad + \beta P_{11ekf}(t)\hat{U}_{ekf}(t) + \beta P_{12ekf}(t)\hat{V}_{ekf}(t)\\ &\quad - G^{-2}P_{11ekf}(t)P_{13ekf}(t))dt,\\ dP_{22ekf}(t) &= (b_2^2 - 2\beta P_{12ekf}(t)\hat{U}_{ekf}(t) - G^{-2}P_{12ekf}^2(t)\\ &\quad - 2\beta P_{22ekf}(t)\hat{V}_{ekf}(t))dt,\\ dP_{23ekf}(t) &= (\beta P_{12ekf}(t)\hat{U}_{ekf}(t) - \delta P_{23ekf}(t)\\ &\quad + \beta P_{22ekf}(t)\hat{V}_{ekf}(t) - \beta P_{23ekf}(t)\hat{V}_{ekf}(t)\\ &\quad - \beta P_{13ekf}(t)\hat{U}_{ekf}(t)\\ &\quad - G^{-2}P_{12ekf}(t)P_{13ekf}(t))dt,\\ dP_{33ekf}(t) &= (2\beta\hat{U}_{ekf}(t)P_{13ekf}(t) + 2\beta\hat{V}_{ekf}(t)P_{23ekf}(t)\\ &\quad - 2\delta P_{33ekf}(t) + b_3^2 - G^{-2}P_{13ekf}^2(t))dt.\end{aligned} \tag{7.38}$$

Numerical simulations for Eqs. (7.29), (7.31), (7.32), (7.37), and (7.38) are performed with the following initial conditions: $\hat{V}(0) = \hat{V}_{ekf}(0) = 1 \times 10^{-2}$, $\hat{U}(0) = \hat{U}_{ekf}(0) = 3.95 \times 10^8$, $\hat{I}(0) = \hat{I}_{ekf}(0) = 0$, $P_{11}(0) = P_{11ekf}(0) = 2$, $P_{12}(0) = P_{12ekf}(0) = 0.5$, $P_{13}(0) = P_{13ekf}(0) = 0.8$, $P_{22}(0) = P_{22ekf}(0) = 2$, $P_{23}(0) = P_{23ekf}(0) = 1$, $P_{33}(0) = P_{33ekf}(0) = 3$.

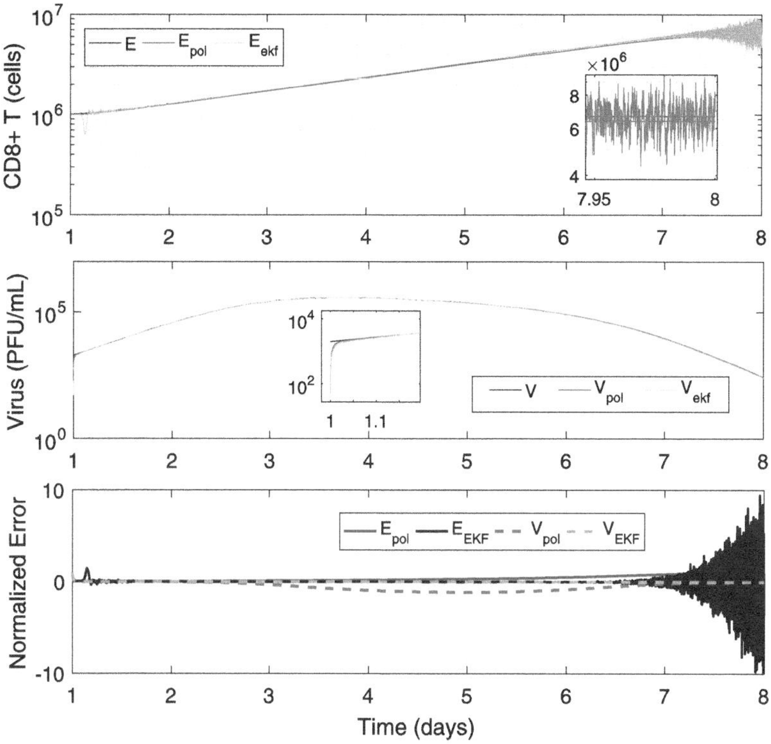

FIGURE 7.1 Model states $V(t)$, $U(t)$, and $I(t)$ with their estimates $\hat{V}(t)$, $\hat{U}(t)$, and $\hat{I}(t)$. Zoom in windows show critical points of performance for both estimators, polynomial and EKF. The bottom panel contains the normalized error of each state for both approaches.

Fig. 7.1 shows the performance of the polynomial and EKF estimators. The three states of the target cell model are plotted and enlarged to allow better presentation for both estimators. In this case, both estimators show similar results in the complete set of time. The fourth panel shows the normalized error estimation for each of the states with both estimators.

7.5.2 Virus immune response model

The virus immune response model [22] involves the action of the CD8+ T cell immune response (E) due to the viral infection (V).

Consider now the following set of equations:

$$
\begin{aligned}
dE(t) &= (S_E + rE(t)\left(\frac{V(t)}{V(t)+k_e}\right) - c_e E(t))dt + dW_E(t),\\
dV(t) &= (pV(t)\left(1-\frac{V(t)}{k_v}\right) - c_v V(t)E(t))dt + dW_V(t),
\end{aligned}
\tag{7.39}
$$

$$dy(t) = V(t)dt + dW_y(t),$$

with initial conditions $E(0) = 10^6$ cells and $V(0) = 25$ PFU/mL and parameter values $r = 0.33$ days^{-1}, $k_e = 2.7 \times 10^3$ PFU/mL, $c_e = 2.0 \times 10^{-2}$ days^{-1}, $S_E = 2.0 \times 10^4$ days^{-1}cells, $p = 4.4$ days^{-1}, $k_v = 10^6$ PFU/mL, and $c_v = 1.24 \times 10^{-6}$ days^{-1}cells^{-1}.

7.5.2.1 Polynomial state estimation

According to Eqs. (7.16), (7.18), (7.20), and (7.21), the filtering equations take the following form:

$$\begin{aligned} d\hat{E}(t) &= (S_E - c_e\hat{E}(t) + r\frac{\hat{E}(t)\hat{V}(t)}{\hat{V}(t)+k_e} + r\frac{P_{12}(t)}{\hat{V}(t)+k_e})dt \\ &\quad + G^{-2}P_{12}(t)(dy(t) - \hat{V}(t)dt), \\ d\hat{V}(t) &= (p\hat{V}(t) - \frac{p}{k_v}\hat{V}^2(t) - c_v\hat{E}(t)\hat{V}(t) - \frac{p}{k_v}P_{22}(t) \\ &\quad - c_vP_{12}(t))dt + G^{-2}P_{22}(t)(dy(t) - \hat{V}(t)dt), \end{aligned} \tag{7.40}$$

$$\begin{aligned} dP_{11}(t) &= (-2c_eP_{11}(t) + \frac{2r\hat{V}(t)P_{11}(t)}{\hat{V}(t)+k_e} + \frac{2r\hat{E}(t)P_{12}(t)}{\hat{V}(t)+k_e} \\ &\quad + b_1^2 - G^{-2}P_{12}^2(t))dt, \\ dP_{12}(t) &= (-c_eP_{12}(t) + pP_{12}(t) \\ &\quad + \frac{r\hat{V}(t)P_{11}(t) + r\hat{V}(t)P_{12}(t) + r\hat{E}(t)P_{22}(t)}{\hat{V}(t)+k_e} \\ &\quad - \frac{p\hat{V}(t)P_{11}(t) + 2p\hat{V}(t)P_{12}(t)}{k_v} \\ &\quad - c_v\hat{V}(t)P_{11}(t) - c_v\hat{E}(t)P_{12}(t) \\ &\quad - G^{-2}P_{12}(t)P_{22}(t))dt, \\ dP_{22}(t) &= (2pP_{22}(t) - \frac{6p\hat{V}(t)P_{22}(t)}{k_v} - 2c_v\hat{V}(t)P_{11}(t) \\ &\quad - 2c_v\hat{V}(t)P_{12}(t) - 2c_v\hat{E}(t)P_{22}(t) \\ &\quad + \frac{2p\hat{V}(t)P_{22}}{k_v} + b_2^2 - G^{-2}P_{22}^2(t))dt. \end{aligned} \tag{7.41}$$

7.5.2.2 Extended Kalman filter

To verify the performance of the proposed filter (7.40)–(7.41), the EKF equations [36] are computed for comparison purposes:

$$d\hat{E}_{ekf}(t) = (S_E + \frac{r\hat{E}_{ekf}(t)\hat{V}_{ekf}(t)}{\hat{V}_{ekf}(t)+k_e} - c_e\hat{E}_{ekf}(t))dt$$

$$
\begin{aligned}
&+ G^{-2}P_{12ekf}(t)(dy(t) - \hat{V}_{ekf}(t)dt), \\
d\hat{V}_{ekf}(t) &= (p\hat{V}_{ekf}(t)(1 - \frac{\hat{V}_{ekf}(t)}{k_v}) - c_v\hat{V}_{ekf}(t)\hat{E}_{ekf}(t))dt \\
&+ G^{-2}P_{12ekf}(t)(dy(t) - \hat{V}_{ekf}(t)dt),
\end{aligned} \tag{7.42}
$$

$$
\begin{aligned}
dP_{11ekf}(t) &= (\frac{2rP_{11ekf}(t)\hat{V}_{ekf}(t)}{\hat{V}_{ekf}(t) + k_e} - 2c_eP_{11ekf}(t) \\
&+ \frac{2rk_e\hat{E}_{ekf}(t)P_{12ekf}(t)}{(\hat{V}_{ekf}(t) + k_e)^2} + b_1^2 \\
&- \frac{P_{12ekf}(t)^2}{G^2})dt, \\
dP_{12ekf}(t) &= (-c_v\hat{V}_{ekf}(t)P_{11ekf}(t) + pP_{12ekf}(t) \\
&- \frac{2p\hat{V}_{ekf}(t)P_{12ekf}(t)}{k_v} - c_v\hat{E}_{ekf}(t)P_{12ekf}(t) \\
&+ \frac{r\hat{V}_{ekf}(t)P_{12ekf}(t)}{\hat{V}_{ekf}(t) + k_e} - c_eP_{12ekf}(t) \\
&+ \frac{rk_e\hat{E}_{ekf}(t)P_{22ekf}(t)}{(\hat{V}_{ekf}(t) + k_e)^2} \\
&- \frac{P_{12ekf}(t)P_{22ekf}(t)}{G^2})dt, \\
dP_{22ekf}(t) &= (-2c_v\hat{V}_{ekf}(t)P_{12ekf}(t) + 2pP_{22ekf}(t) \\
&- \frac{4p\hat{V}_{ekf}(t)P_{22ekf}(t)}{k_v} \\
&- 2c_v\hat{E}_{ekf}(t)P_{22ekf}(t) + b_2^2 - \frac{P_{22ekf}(t)^2}{G^2})dt.
\end{aligned} \tag{7.43}
$$

Numerical simulations using Eqs. (7.39)–(7.43) are carried out with the following initial conditions: $\hat{E}(0) = \hat{E}_{ekf}(0) = 985000$, $\hat{V}(0) = \hat{V}_{ekf}(0) = 50$, $P_{11}(0) = P_{11ekf}(0) = 300$, $P_{12}(0) = P_{12ekf}(0) = 12$, $P_{22}(0) = P_{22ekf}(0) = 120$. Model states, estimates, and the estimation error are presented in Fig. 7.2. The first two simulations display the time evolution of $E(t)$ and $V(t)$, respectively. The third plot corresponds to the time evolution of their corresponding estimation errors.

Results displayed in Fig. 7.2 indicate good estimations for both approaches from the initial time until around day 7; afterwards, the estimation error of EKF highly increases with respect to the polynomial estimation error, especially to the estimation of immune CD8+ T cells. In fact, for the state E, the proposed polynomial method continues to have small er-

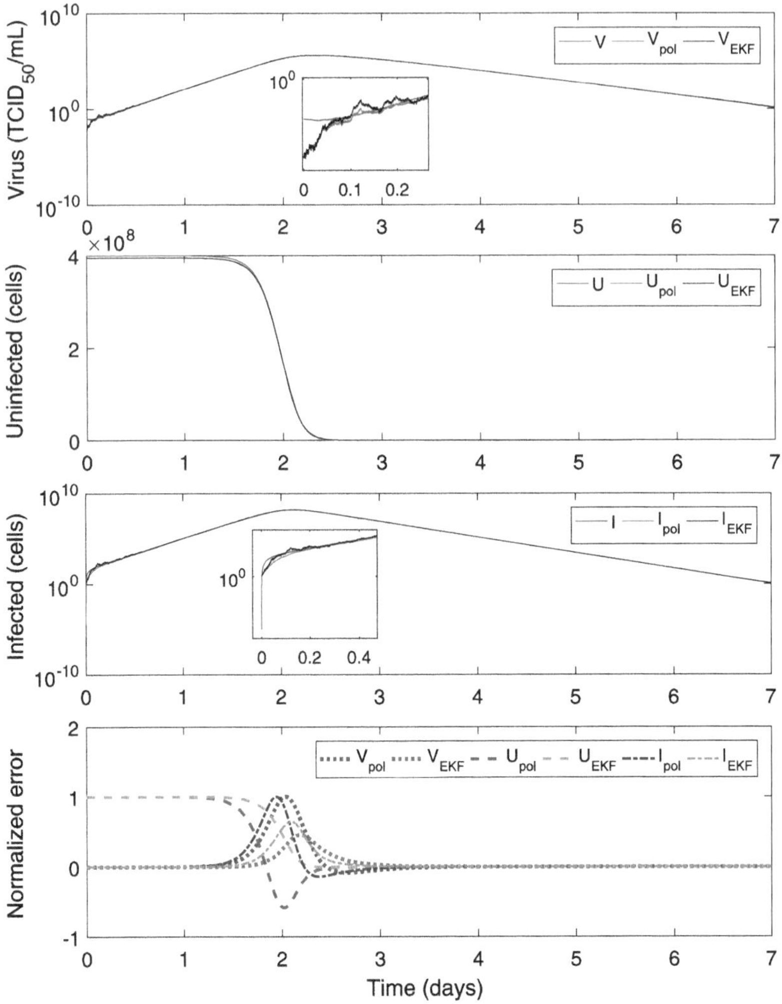

FIGURE 7.2 Model states $V(t)$ and $E(t)$ with their respective estimates. Zoom in windows show critical points of performance for both estimators, polynomial and EKF. The bottom panel contains the normalized error of both approaches.

ror estimations for the complete simulation time, while the EKF error rises higher than 2×10^6.

7.6 Conclusions

The problem of state estimation of stochastic biological systems is presented and developed using polynomial systems. Numerical simulations are used to show the performance for two nonlinear viral infection models involving different processes and dynamics, that is, the target cell system and the virus immune response model.

While both approaches can handle the estimation problem, differences between these techniques are highlighted for Michaelis–Menten terms. The polynomial estimator keeps low error estimation in both models (bilinear and Michaelis–Menten terms) for the complete simulation time. The polynomial approach is a promising tool for future development in clinical decisions and may be considered to expand its use for other biological nonlinear systems. We have also considered the class of polynomial systems where matrix coefficients depend on nonlinear functions.

Future work will consist in merging the observers and impulsive control strategies developed for scheduling therapies [20,21]. Note that for applicability to clinics, several practical limitations need to be resolved since measurements are expensive, infrequent, and with time delays. Thus, machine learning algorithms such as high-order neural networks trained with the cubature KF could help to tackle the model requirements to design observers [18], providing robustness to uncertainties and external disturbances. Even though the design of filters for nonlinear stochastic systems over continuous measurements is an interesting research study, a more realistic application to biological systems occurs in the case of discrete-time measurements, hence research will continue for this class of discrete-time measurement systems [16,28].

References

[1] S.G. Abokhatwa, K.M. El Hadad, M.R. Katebi, Nonlinear state estimation of power plant system using state-dependent Kalman filter, in: 2021 IEEE 1st International Maghreb Meeting of the Conference on Sciences and Techniques of Automatic Control and Computer Engineering MI-STA, 25–27 May, Tripoli, Libya, 2021, pp. 116–121.

[2] H.H. Afshari, S.A. Gadsden, S. Habibi, Gaussian filters for parameter and state estimation: a general review of theory and recent trends, Signal Processing 35 (2017) 218–238.

[3] A. Ahmed, M. Moinuddin, U.M. Al-Saggaf, State space least mean square for state estimation of synchronous motor, in: 8th International Conference on Electrical and Computer Engineering (ICECE), 20–22 Dec., Dhaka, Bangladesh, 2014, pp. 461–464.

[4] A.Y. Alanis, M. Hernandez-Gonzalez, E.A. Hernandez-Vargas, Observers for biological systems, Applied Soft Computing 24 (2014) 1175–1182.

[5] B.D.O. Anderson, J.B. Moore, Optimal Filtering, Dover Publications, Inc., New York, 1979.

[6] P. Baccam, C. Beauchemin, C.A. Macken, F.G. Hayden, A.S. Perelson, Kinetics of influenza A virus infection in humans, Journal of Virology 80 (2006) 7590–7599.

[7] A. Balluchi, L. Benvenuti, M.D. Di Benedetto, The design of dynamical observers for hybrid systems: theory and application to an automotive control problem, Automatica 49 (4) (2013) 915–925.

[8] M.V. Basin, New Trends in Optimal Filtering and Control for Polynomial and Time-Delay Systems, Springer, Berlin, 2008.

[9] M. Basin, D. Calderon-Alvarez, M. Skliar, Optimal filtering for incompletely measured polynomial states over linear observations, International Journal of Adaptive Control and Signal Processing 22 (5) (2008) 482–494.

[10] M. Basin, J. Perez, M. Skliar, Optimal filtering for polynomial system states with polynomial multiplicative noise, International Journal of Robust and Nonlinear Control 16 (6) (2006) 303–314.
[11] S. Bogosyan, M. Barut, M. Gokasan, Braided extended Kalman filters for sensorless estimation in induction motors at high-low/zero speed, IET Control Theory & Applications 1 (4) (2007) 987–998.
[12] C. Briat, M. Khammash, Simple interval observers for linear impulsive systems with applications to sampled-data and switched systems, in: 20th IFAC World Congress, Vol. 50, No. 1, 2017, pp. 5079–5084.
[13] J.B. Burl, Linear Optimal Control H_2 and $\mathcal{H}_\infty$ Methods, Addison-Wesley, Menlo Park, CA, 1999.
[14] H.D. Choi, S.H. You, Fuzzy finite memory state estimation for electro-hydraulic active suspension systems, IEEE Access 9 (2021) 99364–99373.
[15] S. Fakoorian, A. Santamaria-Navarro, B.T. Lopez, D. Simon, A. Agha-mohammadi, Towards robust state estimation by boosting maximum correntropy criterion Kalman filter with adaptive behaviors, IEEE Robotics and Automation Letters 6 (3) (2021).
[16] A. Germani, C. Manes, P. Palumbo, Polynomial extended Kalman filter, IEEE Transactions on Automatic Control 50 (12) (2005) 2059–2064.
[17] S. Julier, J. Uhlmann, H.F. Durrant-Whyte, A new method for the nonlinear transformation of means and covariances in filters and estimators, IEEE Transactions on Automatic Control 40 (3) (2000) 477–482.
[18] M. Hernandez-Gonzalez, E.A. Hernandez-Vargas, M.V. Basin, Discrete-time high order neural network identifier trained with cubature Kalman filter, Neurocomputing 322 (2018) 13–21.
[19] M. Hernandez-Gonzalez, M.V. Basin, Discrete-time filtering for nonlinear polynomial systems over linear observations, International Journal of Systems Science 45 (7) (2014).
[20] G. Hernandez-Mejia, A.Y. Alanis, E.A. Hernandez-Vargas, Neural inverse optimal control for discrete-time impulsive systems, Neurocomputing 314 (2018) 101–108.
[21] G. Hernandez-Mejia, A.Y. Alanis, M. Hernandez-Gonzalez, R. Findeisen, E.A. Hernandez-Vargas, Passivity-based inverse optimal impulsive control for influenza treatment in the host, IEEE Transactions on Control Systems Technology (2019).
[22] E.A. Hernandez-Vargas, Modeling and Control of Infectious Diseases: With MATLAB and R, 1st edition, Elsevier Academic Press, 2019.
[23] R.E. Kalman, A new approach to linear filtering and prediction problems, Transactions of the ASME. Series D, Journal of Basic Engineering 82 (1960) 34–45.
[24] R.E. Kalman, R.S. Bucy, New results in linear filtering and prediction theory, Transactions of the ASME. Series D, Journal of Basic Engineering 83 (1961) 95–107.
[25] Y. Kobayashi, I. Ono, Sequential estimation of states and parameters of nonlinear state space models using particle filter and natural evolution strategy, in: 2020 IEEE Congress on Evolutionary Computation (CEC), 19–24 July, Glasgow, 2020, pp. 1–8.
[26] M. Lefebvre, Applied Stochastic Process, Springer, USA, 2007.
[27] T. Liu, X. Zhang, S. Zhu, Study of state estimation with super particle, in: 2006 6th World Congress on Intelligent Control and Automation, 21–23 June, Dalian, China, 2006, pp. 1434–1437.
[28] M.B. Luca, S. Azou, G. Burel, A. Serbanescu, On exact Kalman filtering of polynomial systems, IEEE Transactions on Circuits and Systems. I, Regular Papers 53 (6) (2006) 1329–1340.
[29] S. Moontaha, A. Galka, T. Meurer, M. Siniatchkin, Analysis of the effects of medication for the treatment of epilepsy by ensemble iterative extended Kalman filtering, in: 2018 40th Annual International Conference of the IEEE Engineering in Medicine and Biology Society (EMBC), 18–21 July, Honolulu, HI, USA, 2018, pp. 187–190.
[30] T. Ohhira, A. Shimada, T. Murakami, Variable forgetting factor-based adaptive Kalman filter with disturbance estimation considering observation noise reduction, IEEE Access 9 (2021) 100747–100756.

[31] S.C. Patwardhan, S. Narasimhan, P. Jagadeesan, Nonlinear Bayesian state estimation: a review of recent developments, Control Engineering Practice 20 (2012) 933–953.
[32] V.S. Pugachev, Probability Theory and Mathematical Statistics for Engineers, Pergamon Press, United Kingdom, 1984.
[33] V.S. Pugachev, I.N. Sinitsyn, Stochastic Systems. Theory and Applications, World Scientific, Singapure, 2001.
[34] A. Sagan, Y. Liu, A. Bernstein, Decentralized low-rank state estimation for power distribution systems, IEEE Transactions on Smart Grid 12 (4) (2021) 3097–3106.
[35] X. Shen, L. Deng, A dynamic system approach to speech enhancement using the $\mathcal{H}_\infty$ filtering algorithm, IEEE Transactions on Speech and Audio Processing 7 (4) (1999) 391–399.
[36] D. Simon, Optimal State Estimation. Kalman, H-Infinity, and Nonlinear Approaches, John Wiley & Sons, New Jersey, 1999.
[37] L. Szilagyi, Application of the Kalman filter in cardiac arrhythmia detection, in: Proceedings of the 20th Annual International Conference of the IEEE Engineering in Medicine and Biology, Vol. 20, No. 1, 1998, pp. 98–100.
[38] T.-J. Tarn, Y. Rasis, Observers for nonlinear stochastic systems, IEEE Transactions on Automatic Control AC-21 (4) (1976) 441–448.
[39] T. Tayama, Y. Kurose, M.M. Marinho, K. Harada, S. Omata, F. Arai, K. Sugimoto, F. Araki, K. Totsuka, M. Takao, M. Aihara, M. Mitsuishi, Autonomous positioning of eye surgical robot using the tool shadow and Kalman filtering, in: 2018 40th Annual International Conference of the IEEE Engineering in Medicine and Biology Society (EMBC), 18-21 July, Honolulu, HI, USA, 2018, pp. 1723–1726.
[40] Y. Wang, L. Xu, F. Zhang, H. Dong, Y. Liu, G. Yin, An adaptive fault-tolerant EKF for vehicle state estimation with partial missing measurements, IEEE/ASME Transactions on Mechatronics 26 (3) (2021).
[41] W.H.O., The World Health Organization Influenza (Seasonal) Fact Sheet N. 211, World Health Organization, 2017.
[42] Z. Zhang, K.H. Wong, A computer vision based sea search method using Kalman filter and CAMSHIFT, in: International Conference on Technological Advances in Electrical, Electronics and Computer Engineering (TAEECE), Konya, Turkey, 2013, pp. 188–193.
[43] J. Zhao, A. Gómez-Expósito, M. Netto, L. Mili, A. Abur, V. Terzija, I. Kamwa, B. Pal, A.K. Singh, J. Qi, Z. Huang, A.P. Sakis Meliopoulos, Power system dynamic state estimation: motivations, definitions, methodologies, and future work, IEEE Transactions on Power Systems 34 (4) (2019) 3188–3198.
[44] R. Zurakowski, Nonlinear observer output-feedback MPC treatment scheduling for HIV, Biomedical Engineering Online. BioMed Central 10 (40) (2011) 1–16.

CHAPTER

8

Sliding mode control theory interprets elite control of HIV

Anet J.N. Anelone[a], *Peter Kim*[a], *and Sarah K. Spurgeon*[b]

[a]School of Mathematics and Statistics, University of Sydney, Camperdown, NSW, Australia, [b]Department of Electronic and Electrical Engineering, University College London, London, United Kingdom

8.1 Introduction

The motivation for this chapter comes from the imperative to advance the understanding of virus control by the immune system. In particular, it is important to understand how to sustain human immunodeficiency virus (HIV) suppression in the absence of antiretroviral therapy to yield a functional cure for HIV [1]. HIV type 1 (HIV-1) infection leads to high viral load (RNA copies per mL) and low CD4+ T cell counts (cells/mm^3) in the peripheral blood of most untreated individuals [2,3,1]. In 2005, clinical studies showed that less than 1% of individuals infected with HIV-1 spontaneously control and maintain the viral load below detectable levels, i.e., <50 RNA copies per mL, over time [4]. This rare phenomenon is known as elite control of HIV [5,2,1]. Individuals exhibiting elite control tend to be infected with pathogenic and replication-competent virus [5,4,2], yet they typically suppress the viral load to undetectable levels early (approximately within 6 months) following primary infection [5,6]. They also tend to have protective human leukocyte antigen (HLA) alleles and qualitatively better HIV-specific CD8+ T cell responses with higher killing efficacy than individuals who experience a progression of HIV infection [5,2].

Undetectable HIV loads are desirable for several clinical reasons such as a better quality of life and to avoid acquired immunodeficiency syndrome

Feedback Control for Personalized Medicine
https://doi.org/10.1016/B978-0-32-390171-0.00017-2

(AIDS) [4]. Numerous studies have been conducted to elucidate the mechanism involved in elite control of HIV [5,2], but such studies have been limited by the lack of reliable animal models [5,2].

In addition to studies using animal models, HIV infection has also been studied using mathematical modeling of the within-host dynamics of the virus [7,3]. Indeed, mathematical modeling has provided significant qualitative and quantitative insights on the pathogenesis and kinetics of HIV and on the immune mechanisms involved in viral control [7,8,3]. Seminal work modeling the changes over time of HIV RNA levels and CD4+ T cell counts during antiretroviral therapy has profoundly changed understanding of HIV dynamics by demonstrating the rapid turnover of HIV in vivo [3,8,9]. Mathematical models of the within-host dynamics of HIV are often constructed using ordinary differential equations (ODEs). Although the immune response influences the progression of HIV, the basic target cell limited model does not include any immune response yet provides a reasonable fit to available clinical data on the acute phase of HIV infection [10,3]. The realism of the basic HIV model has been improved by including relevant cell populations such as latently infected cells, macrophages, and HIV-specific CD8+ T cells [3,7,11,12]. Elite control of HIV is simulated using a high killing rate of HIV-specific CD8+ T cells to drop viral load to undetectable levels [11]. The modeling work in [12] formulated the first ODEs to describe acute HIV infection and the progression to AIDS. This model introduced an additional feedback for HIV pathogenesis, where the slow infection of macrophages by HIV triggers the rise of the viral load in the last stages of the infection and the subsequent slow depletion of CD4+T cells [12].

Additional investigations are required to improve understanding of currently uncertain biological rates and processes leading to elite control [11,7,3]. Since the killing process of CD8+ T cells and antiretroviral treatments are suitable means to change the viral load in finite time, this matches the notion of *controllability* [13,14]. Controllability refers to the ability of a given input to steer a system from any initial state to another state in finite time [13,14]. Therefore, it is natural to apply control theory to consider in a particular system robustness because the field is a mature discipline in the study of uncertain and nonlinear dynamical processes which are subject to modification via feedback [15,16].

Previous control engineering studies have analyzed HIV infection dynamics in vivo as a closed-loop system in which therapies are control inputs [17–19]. Control theory is applied to suggest dosages and drug schedules to induce clinical outcomes better than those obtained by standard clinical treatments. HIV infection during antiretroviral therapy has also been studied as an impulsive control system (ICS), in which the control impulse represents the administration of antiretroviral drugs, since this control input has a short duration in comparison with the sampling time and the dynamics of the infection [20,19]. For this ICS, the free response corresponds to the time evolution of the states when the treatment is not administered, and the forced response corresponds to instantaneous jumps when treatments are administered [20,19]. This approach has been used to propose treatment protocols which enhance the immune responses against HIV [21]. Strategies involving optimal control and model predictive control (MPC) have also been designed to improve the suppression of viral load during therapy and to determine suitable scheduling for treatment interruptions [22–25]. MPC strategies have been investigated to suggest optimal treatment scheduling to reduce HIV escape mutations [17]. MPC is particularly useful to consider input constraints related to the efficacy and safety of drugs and state constraints related to the therapeutic goals and clinical constraints. Since latently infected cells are a major hurdle to achieve HIV cure, the work in [26] developed an optimal control strategy which includes the infusion of HIV antigen as a second control input. This strategy leads to a reduction of the population of latently infected cells and allows the viral load to remain within a desired therapeutic range after treatment interruption [26].

The results in [18,27,28] have demonstrated the synergies between the dynamics of the immune response of CD8+ T cells and the dynamics of switched control strategies such as *sliding mode control*. As in the immune system, a sliding mode control system is characterized by a set of feedback loops and a decision rule that switches the applied control to achieve desired dynamical behavior [28,15,16]. The desired dynamical behavior is used to formulate a switching function $s(t)$, which is a function of the state variables of the system. The switching function $s(t)$ is formulated such that when an ideal sliding motion occurs on the manifold $s(t) = 0$, the desired dynamical behavior is achieved. Consider the following Lyapunov function candidate for the sliding mode dynamics, where $\vee = 0.5s^2$ and $\frac{d\vee}{dt} = s(t)\frac{s(t)}{dt}$. An ideal sliding motion is enforced when a control is selected so that the so-called *reachability paradigm* $s(t)\frac{s(t)}{dt} < 0$ is satisfied. The reachability paradigm is a sufficient condition for the attainment of the sliding mode in finite time. Typically, a sliding mode controller is designed using a discontinuous or a saturation function, and the control gain is tuned to ensure that the reachability condition is satisfied at all times. This is important and advantageous for several applications because when the reachability condition holds and $s(t) = \dot{s}(t) = 0$, the effective action of the applied sliding mode controller, i.e., the *equivalent control* action, enforces complete robustness to disturbances and uncertainty implicit in the input channel. Note that the equivalent control is not the actual control signal applied to the system; the equivalent control represents the theoretical control action to maintain $s(t) = \dot{s}(t) = 0$. Thus, sliding mode control theory has the potential to provide a sensible theoretical and analytical framework to study immunological control mechanisms.

In this chapter, unlike in typical control engineering studies, a controller is not sought, nor implemented. Instead, control theory is applied to deliver an analytic and quantitative framework to study viral control, which is assumed to be an inherent feedback control mechanism, in vivo. Here a control analysis is performed to understand the control laws and robustness properties of elite control of HIV. It is hypothesized that the controller is the cytolytic killing mechanism of the specific CD8+ T cell response. The approach used in this chapter is novel because the reachability paradigm, which is typically used to assess the reachability and robustness of a sliding mode control [15,16], is applied to study elite control of HIV. A crucial challenge in the design of sliding mode controllers for engineering systems is the formulation of the switching function which prescribes the ideal system dynamics in the sliding mode. Here, the switching function is inherent in the observed immune response. The main contribution of this work is to apply sliding mode control theory to analyze elite control of HIV; the reachability paradigm analytically determines T cell immunity and explains the robustness of elite control of HIV to uncertainties and perturbations in biological rates and mechanisms. Together, the findings suggest that the study of virus dynamics could benefit from the application of mature analytical tools from sliding mode control.

The chapter is organized as follows. Section 8.2 presents a model of within-host HIV infection that incorporates uncertainties; Section 8.3 presents an analytical analysis of elite control of HIV using the reachability paradigm; Section 8.4 highlights the robustness properties of elite control of HIV infection; and Section 8.5 discusses simulation of immunological scenarios that are conducted to show that the reachability paradigm appropriately predicts and monitors the robustness of elite control of HIV.

8.2 Modeling HIV infection in vivo

The dynamical equations used to model HIV within-host dynamics are inspired from recent models in the literature [11,7,8,3]. The dynamical equations of the model are written to encompass modeling uncertainties in cell growth dynamics, in the activation or inhibition of the immune response, and in the killing process of effector cells. The model describes the infection of target cells, T, by free HIV virions, V. The function $\beta(t)$ describes the variation over time of the HIV infection rate. The function $g_T(T)$ describes the growth dynamics of target cells. A proportion $\alpha_L(t)$ of target cells become latently infected cells, L. The growth dynamics of latently infected cells are described by the function $g_L(L)$. Latently infected cells become productively infected cells, I, at rate $w_L(t)$. Productively infected cells die at rate δ and produce new free HIV virions at rate p. Free

HIV virions are cleared in the blood stream at rate c. Productively infected cells trigger the activation of HIV-specific effector CD8+ T cells, E, and this process is described by the function $f(I, E)$. HIV infection also inhibits immune activation and this is described by the function $d(I, E)$). The growth dynamics of effector cells are described by the function $g_E(E)$. Effector cells kill productively infected cells and this killing process is described by the function $u_C(I, E)$. This chapter studies the killing process of productively infected cells by effector cells as an intrinsic feedback control that enforces elite control of HIV infection. Hence, the system of ODEs modeling HIV within-host dynamics is written as follows:

$$\begin{aligned}
\frac{dT}{dt} &= g_T(T) - \beta(t)TV, \\
\frac{dL}{dt} &= \alpha_L(t)\beta(t)TV - w_L(t)L + g_L(L), \\
\frac{dI}{dt} &= (1 - \alpha_L(t))\beta(t)TV - \delta I + w_L(t)L - u_C(I, E), \\
\frac{dV}{dt} &= pI - cV, \\
\frac{dE}{dt} &= g_E(E) + f(E, I) - d(E, I).
\end{aligned} \tag{8.1}$$

The state variables are target cells, T, latently infected cells, L, productively infected cells, I, free HIV virions, V, and effector cells, E. The model (8.1) generalizes and unifies previous models from [11,7,8,3]. This approach was followed because the models simulate the sophisticated dynamical interactions between HIV and immune cells well with relatively few and simple equations. The model (8.1) was formulated to encompass nonlinearities as well as some parameter and modeling uncertainties in the within-host dynamics of HIV infection. It is important to deliver robust predictions because parameter variations within a host and between hosts are expected. Some parameters may also vary during the course of infection, and the functions modeling some biological processes are the subject of active experimental and mathematical validation [7]. The control analysis of this model aims to deliver robust predictions, since sliding mode control theory is a robust control paradigm [15].

The nominal model of (8.1) is represented in Fig. 8.1 and configured using the functions in [11].

In this nominal model, the growth function of target cells is

$$g_T(T) = \lambda - dT. \tag{8.2}$$

The growth function of latently infected cells is

$$g_L(L) = (\rho - d_L)L. \tag{8.3}$$

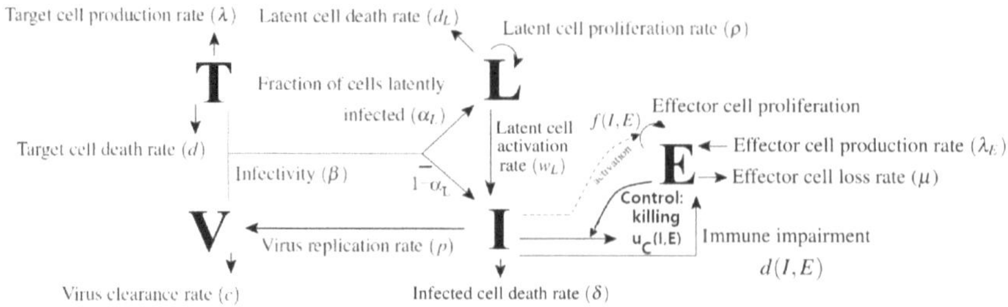

FIGURE 8.1 Diagram of the model (8.1) describing HIV infection dynamics in vivo. T, target cells; L, latently infected cells; I, productively infected cells; V, free HIV virions; and E, effector cells. The biological rates for this nominal model are given in Table 8.1.

The growth function of effector cells is

$$g_E(E) = (\lambda_E - \mu)E. \tag{8.4}$$

The activation function of effector cells is

$$f(I, E) = b_E \frac{I}{K_B + I} E. \tag{8.5}$$

The function modeling the inhibition of effector cells is

$$d(I, E) = d_E \frac{I}{K_D + I} E. \tag{8.6}$$

The candidate function modeling the intrinsic control of the killing process effector cells is

$$u_C(I, E) = mIE, \tag{8.7}$$

where $0 \leq m \leq 1$ (mL/cells/day) is the killing rate of effector cells. This killing rate can be viewed as a high gain control action applied by the immune response in the context of elite control of HIV because HIV-specific CD8+ T cells tend to have a high killing efficacy in elite HIV controllers [5].

Elite control of HIV is simulated using the nominal model in Fig. 8.2 and the biological rates in Table 8.1. Upon infection on day 0, the population of susceptible target cells T decreases, and a fraction of the target cells are latently infected. The population of latently infected cells decreases slowly over time, since these cells die at a slow rate, and a small number of latently infected cells become virus-producing infected cells. Concurrently, HIV infection triggers clonal expansion of HIV-specific effector cells. Effector cells produce a potent killing action which reduces the population of infected cells continuously, and the viral load declines to undetectable

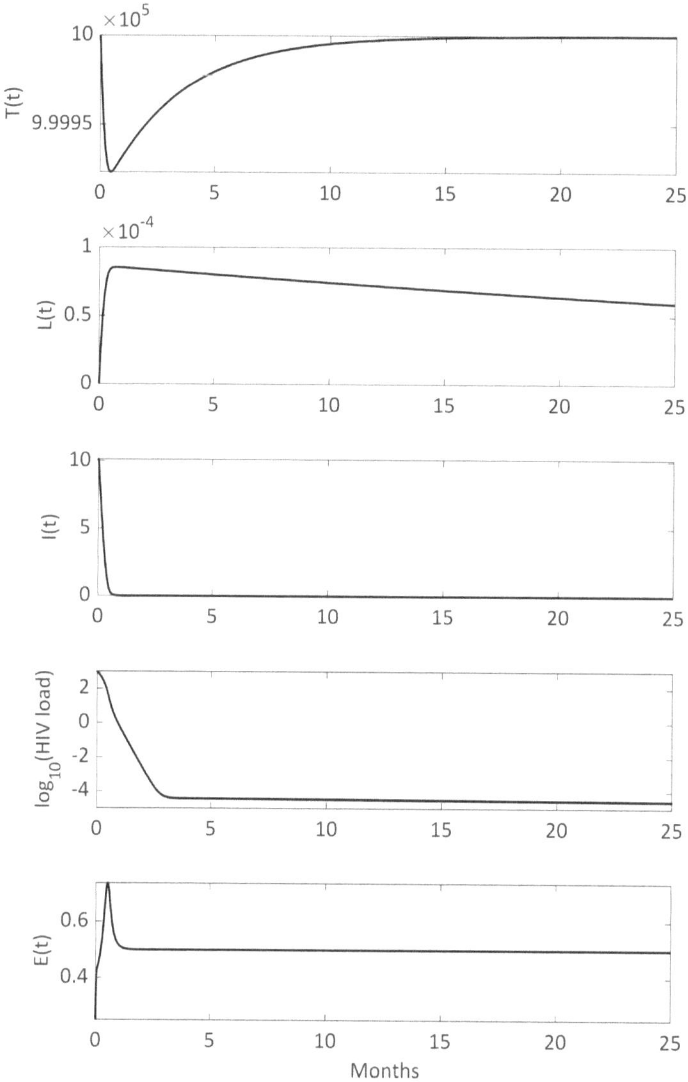

FIGURE 8.2 Simulations of elite control of HIV infection using the nominal model. Initial conditions: target cells $T(0) = 10^6$; latently infected cells $L(0) = 0$; productively infected cells $I(0) = 10$; viral load $V(0) = 1000$; effector cells $E(0) = 0.25$.

levels within 6 months postinfection, as observed in reported cases of elite control of HIV [5,2,1]. This nominal model will be used as a platform to conduct reachability and robustness analyses for elite control of HIV.

TABLE 8.1 Biological rates of the nominal model.

Description (symbol)	Value [reference]
Target cell production rate (λ)	10^4 cells/mL/day [29]
Target cell death rate (d)	0.01 day^{-1} [30]
Mass action infectivity (β)	$1.5 * 10^{-8}$ mL/day [10]
Fraction of latently infected cells (α_L)	10^{-6} [11]
Latent cell proliferation rate (ρ)	0.0045 day^{-1} [31]
Latent cell activation rate (w_L)	0.001 day^{-1} [11]
Latent cell death rate (d_L)	0.004 day^{-1} [32]
Infected cell death rate (δ)	1 day^{-1} [33]
Effector cell killing rate (m)	0.95 mL/cells/day [11]
Virus replication rate (p)	2000 day^{-1} [11]
Virus clearance rate (c)	23 day^{-1} [34]
Effector cell production rate (λ_E)	1 cell/mL/day [11]
Effector cell proliferation coefficient (b_E)	1 day^{-1} [35]
Effector cell production Hill function scaling (K_B)	0.1 cells/mL [36]
Saturation of immune impairment coefficient (d_E)	2 day^{-1} [37]
Saturation of immune impairment Hill function scaling (K_D)	5 cells/mL [36]
Effector cell loss rate (μ)	2 day^{-1} [11]

8.3 A reachability condition for elite control of HIV infection

The reachability paradigm from sliding mode control theory is applied to test different assumptions on the governing dynamics and control objective of elite HIV control, since different sliding manifolds can be formulated to describe different processes to achieve elite HIV control. In particular, the reachability paradigm is applied to understand whether elite HIV control requires an ongoing control action and whether the control action required to achieve the initial viral suppression is similar to or distinct from the control action required to maintain long-term viral suppression [1]. This chapter proposes two candidate sliding manifolds. Importantly, the sliding manifolds are designed to be consistent with immunology. The design process initially considers the characteristics of the dynamic behavior of elite control of HIV infection to formulate a candidate sliding manifold, so that when the solutions of the system (8.1) lie on the chosen manifold, the dynamics of elite control of HIV are exhibited by (8.1). First a manifold is defined based on the philosophy that elite control of HIV is a destination and the corresponding immunological control objective is to reduce the population of productively infected cells to zero. In the second case, a manifold is designed with the philosophy that elite control of HIV is an ongoing process and the corresponding immunological control objective is to decrease the viral load.

8.3.1 Reachability analysis for reducing the population of productively infected cells to zero

In this analysis, the sliding surface is selected as

$$s_E(t) = I, \tag{8.8}$$

which is nonnegative at all times because it represents the number of productively HIV-infected cells. This sliding surface is considered because the sliding manifold $s_E(t) = I = 0$ satisfies the control objective to reduce the population of productively infected cells to zero so as to exhibit the dynamics of elite control of HIV. In fact, when an ideal sliding mode takes place, i.e., $s_E(t) = I = 0$, the viral load $V \to 0$ and HIV is contained.

According to the theory of sliding mode control [15,16], the reachability condition to enforce ideal sliding motion is

$$\begin{aligned} &s_E \frac{ds_E}{dt} < 0, \\ &(1 - \alpha_L(t))\beta(t)TV - \delta I + w_L(t)L - u_C(I, E) < 0, \end{aligned} \tag{8.9}$$

since $0 \leq I$ by definition.

The reachability condition (8.9) suggests that the population of virus-producing infected cells must decline continuously to enforce and sustain viral clearance. Eq. (8.9) suggests that immune responses against HIV continuously reduce the number of virus-producing infected cells in the context of elite control of HIV. Elite control of HIV can thus be obtained due to one or multiple types of immune responses. This helps to understand the heterogeneity in the immunological mechanisms which tend to enforce elite control of HIV [5,6,4,2]. For instance, the killing efficacy of effector cells $u_C(I, E)$ could be high enough on its own or this killing action could be supported by antibody responses against HIV that tend to reduce the infection rate $\beta(t)$. The reachability condition (8.9) also suggests that the requirements for elite control of HIV vary during the progression of the infection due to changes in the biological rates, in the cell population, and in the viral load.

8.3.2 Reachability analysis for decreasing the viral load continuously

An alternative sliding manifold is designed by considering that elite control of HIV ensures the decline of the viral load. The proposed sliding surface is given as

$$s_V(t) = rV + \frac{dV}{dt}, \tag{8.10}$$

where r is a positive scalar.

The corresponding reachability condition is

$$s_V \frac{ds_V}{dt} < 0,$$
$$(rV + pI - cV)(r\frac{dV}{dt} + p\frac{dI}{dt} - c\frac{dV}{dt}) < 0. \tag{8.11}$$

When this fundamental criterion is satisfied, the solutions of the system are confined to the sliding manifold $s_V(t) = 0$. Consequently, the viral load declines at a rate r, i.e., $\frac{dV}{dt} = -rV$. Thus, the dynamics of the model (8.1) on the manifold $s_V(t) = 0$ are consistent with the decline of the viral load to undetectable levels as observed in elite control of HIV [5,11,6]. In particular, when $r = c$, ideal sliding motion is consistent with the decline of the viral load at the natural clearance rate. In this configuration, the reachability condition (8.11) becomes

$$p^2 I \frac{dI}{dt} < 0,$$
$$\frac{dI}{dt} = \frac{ds_E}{dt} < 0, \tag{8.12}$$

since $s_E(t) = I$, $p > 0$, and $I \geq 0$. Therefore, the reachability condition (8.12) is identical to (8.9). Thus, elite control of HIV yields a sliding mode on the manifolds (8.8) and (8.10) which prescribe viral clearance. Together, these reachability analyses suggest that sliding mode control theory interprets elite control of HIV.

8.4 Robustness properties of elite control of HIV infection

The robustness of a sliding mode controller results from the controller's ability to satisfy the reachability condition and to enforce the sliding motion in the presence of uncertainty [15]. The reachability condition provides a nonlinear means to evaluate the robustness of any sliding mode dynamics. Considering (8.9) and (8.12) in the same context is a useful approach to evaluate the robustness properties of elite control of HIV using the well-established framework of sliding mode control theory [27,28,15,16]. The reachability conditions (8.9) and (8.12) capture fundamental and time-dependent criteria for viral clearance, since the design of the sliding manifolds is guided by immunology and the chosen HIV model reasonably well describes complex biological and immunological mechanisms of HIV infection with relatively simple dynamical equations. The approach enables the impact of variations in biological rates and modeling uncertainties on the dynamics of elite control of HIV to be formally explored.

Sliding mode control theory classifies as matched uncertainties those uncertainties implicit in the input channels of the control signals [15]. When this is not the case, the uncertainty is called unmatched uncertainty. Importantly, an appropriate sliding mode control completely rejects the effects of matched uncertainties when the control gain satisfies the reachability condition and ideal sliding motion takes place [15]. In the current study of elite control of HIV using the model (8.1), $u_C(I, E)$ operates as an intrinsic immunological control within the third state equation relating to the population dynamics of productively infected cells, I. Therefore, parameter and modeling uncertainties implicit in this channel are classified as matched uncertainties. Thus, the dynamics of elite control of HIV remain sensitive to unmatched uncertainty such as the growth dynamics of cells, the activation and proliferation of effector cells, and the inhibition of effector cells by HIV. However, as long as the killing action is sufficient to satisfy the reachability condition and maintain ideal sliding motion on $s_E(t) = 0$ or $s_V(t) = 0$, the dynamics of elite control of HIV are robust to matched parameter and modeling uncertainty such as $\beta(t)$, the infection rate of HIV; $w_L(t)L$, the activation of latently infected cells; and unknown biological perturbations. This robustness property of elite control of HIV is achievable by any candidate functions modeling the killing process of effector cells where the output of the control is sufficiently large to satisfy the reachability condition and maintain ideal sliding motion despite the presence of perturbations.

Since the inherent robustness of sliding mode control is typically demonstrated by the principle of the *equivalent control* [15,16], the principle of the equivalent injection is used to interpret the robustness of elite HIV control. Assume the reachability condition $s_E(t) = 0$ holds and differentiate (8.8) with respect to time so that

$$\dot{s}_E(t) = (1 - \alpha_L(t))\beta(t)TV - \delta I + w_L(t)L - u_{eqE}(t)\dot{s}_E(t) = 0.$$

Here

$$u_{eqE}(t) = (1 - \alpha_L(t))\beta(t)TV - \delta I + w_L(t)L$$

is the equivalent control, i.e., the theoretical control action required to maintain the sliding motion. When $s_V(t) = \dot{s}_V(t) = 0$, the equivalent control is given by

$$\begin{aligned} u_{eqV}(t) = (r - c)\frac{dV}{dt} + (1 - \alpha_L(t))\beta(t)TV \\ - \delta I + w_L(t)L. \end{aligned}$$

It is clear that when $r = c$, $u_{eqV}(t) = u_{eqE}(t)$. Since $u_{eqE}(t)$ and u_{eqV} represent the magnitude of cytolytic killing by elite HIV-specific CD8+ T cells, the values of $u_{eqE}(t)$ and u_{eqV} are nonnegative by definition. The theoretical control (cytolytic killing) to maintain $s_V(t) = 0$, i.e., to maintain the decline of the viral load at the natural clearance rate c, is identical to the one required to maintain $s_E(t) = 0$ and so kill all productively infected cells. Hence, the equivalent control can be used as a tool to evaluate the theoretical killing efficacy to maintain elite control of HIV in vivo. Although this work does not implement nor design any biological controller, the above analytical findings suggest that sliding mode control theory can explain both the control mechanism and the robustness of elite control of HIV. Next, immunological experiments are simulated to support the analytical findings above.

8.5 Simulation results

First, the dynamics of elite control of HIV infection in vivo are simulated using the nominal model to demonstrate that the proposed reachability analysis appropriately captures the required dynamics. The elite HIV-specific CD8+ T cell response is seen to suppress the viral load. The proposed reachability condition (8.9) is satisfied and viral clearance is consistent with the attainment of a sliding mode on the manifold $s_E(t) = 0$ and $s_V(t) = 0$ (see Fig. 8.3). This validates the assertion that the reachability analysis developed in this chapter is an appropriate condition to evaluate elite control.

Next, the sliding mode control framework is used to address the question of whether the immune mechanisms that maintain elite HIV control are the same as those that initially brought elite HIV control [1]. To this end, the impact of variations in the killing rate is assessed. As expected, when the killing rate increases, the viral load decreases (see Fig. 8.4). Though the control of HIV is more potent, the magnitude of the killing action decreases, as the viral load and the number of infected cells decrease (see Figs. 8.3 and 8.4). These results are consistent with clinical findings showing that women tend to have lower viral load set points as compared to men during elite control of HIV, since anti-HIV responses in women tend to be more potent than the ones in men [38]. When the killing rate increases, the equivalent control action tends to decrease (see Fig. 8.4), suggesting a reduction in the killing action required to maintain elite control of HIV. This supports the notion that it is detrimental to maintain very potent immune responses in the long term after elite control is achieved, because potent immune responses enhance chronic inflammation [1]. In the simulations in Fig. 8.4, the control input and the equivalent control become low during elite control of HIV, due to the reduction of the viral load and the numbers of latently infected cells. These results support the notion that for elite control of HIV, getting there is not the same as staying

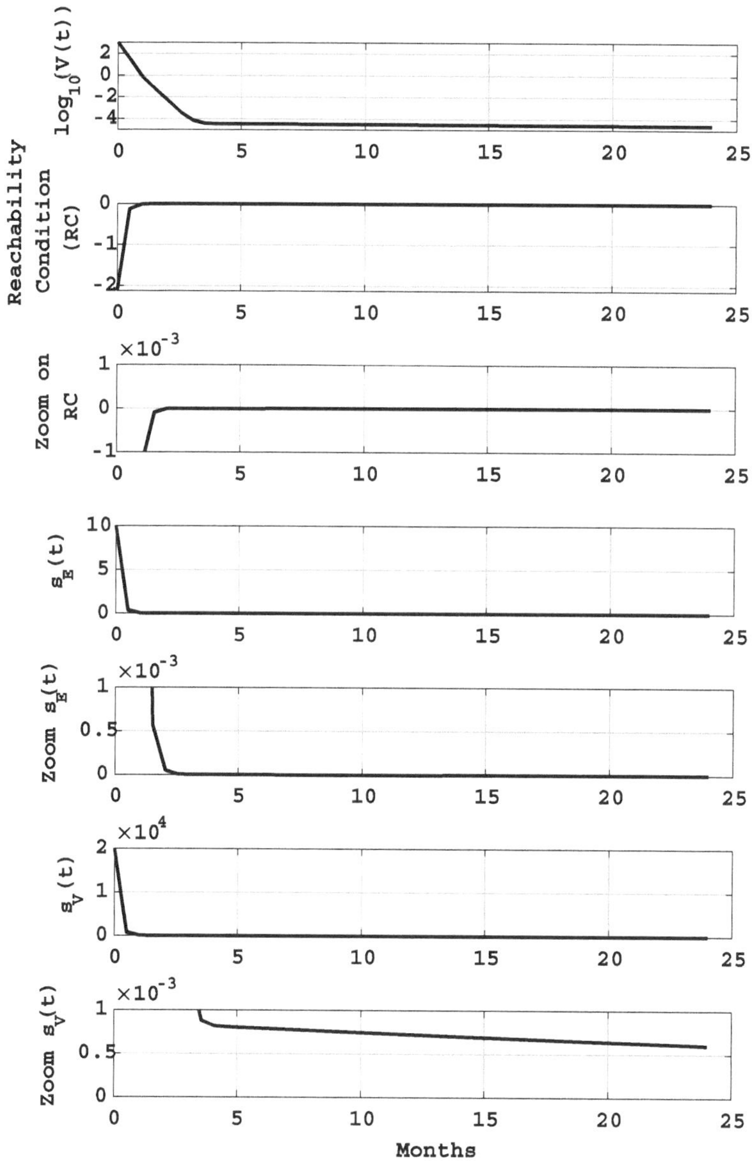

FIGURE 8.3 Sliding mode control theory interprets elite control of HIV. Simulation of the nominal model with nominal parameter values and initial conditions: $T(0) = 10^6$; $L(0) = 0$; $I(0) = 10$; $V(0) = 1000$; $E(0) = 0.75$.

there; potent control of HIV is required to achieve elite control, but afterwards, less potent immune responses, which yield less inflammation, are desirable to maintain long-term viral suppression [1]. Interestingly, in en-

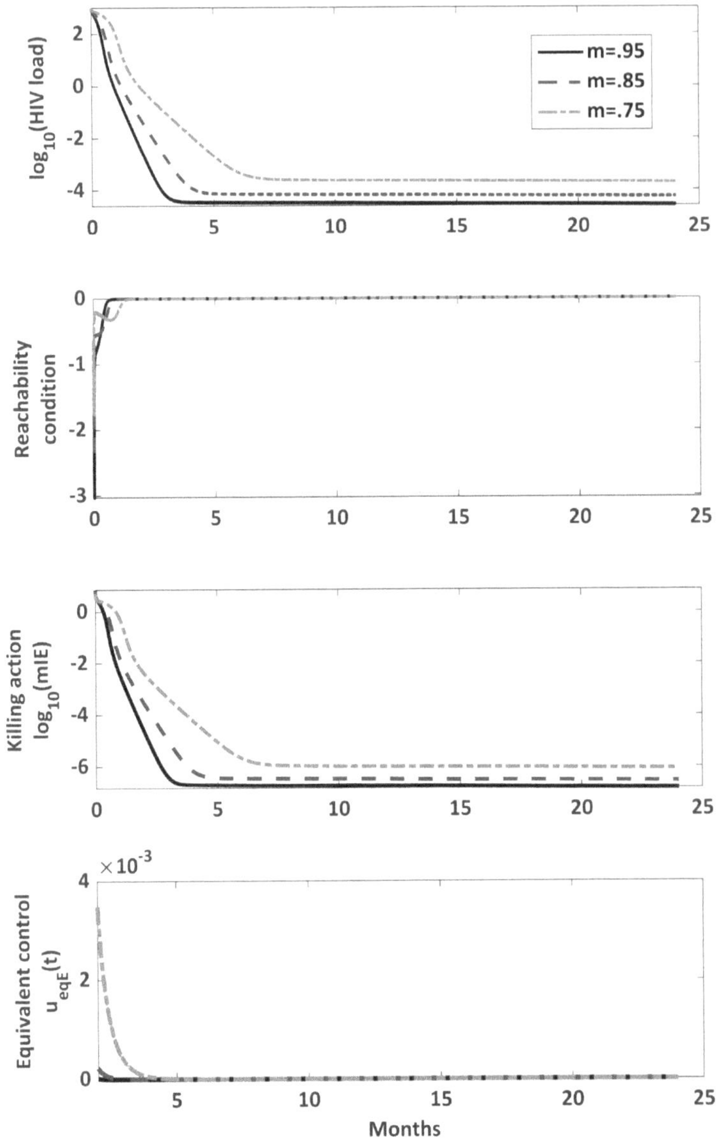

FIGURE 8.4 Simulations of elite control of HIV with variations in the killing rate m. Initial conditions: $T(0) = 10^6$; $L(0) = 0$; $I(0) = 10$; $V(0) = 1000$; $E(0) = 0.75$.

gineering systems with a sliding mode control, the equivalent control also tends to decrease during the sliding motion. The design of sliding mode controllers often incorporates adaptation schemes to reduce the controller effort and to improve performance. This is consistent with experimental studies which observed that the activity of HIV-specific immune responses are low and barely detectable even though elite control of HIV is exhibited [1]. Since high chronic inflammation tends to contribute to viral rebound and/or impair the immune system [1], elite control of HIV seems to involve a natural adaptation scheme to reduce the killing action and inflammation, such that viral control becomes more likely to be maintained once it is achieved.

The simulation results presented in Fig. 8.5 represent an experimental study in which CD8+ T cells are depleted to test whether failure of elite control of HIV is consistent with failure of the sliding mode. This depletion is modeled by inducing the death of effector cells with the term $-3E$ after 6 months in the nominal model of elite control of HIV. Therefore, this depletion experiment introduces an unmatched biological perturbation for elite control of HIV because this perturbation is not implicit in the input channel of the killing process. As expected, the number of effector cells drops drastically after the start of the depletion process and remains negligible during the depletion period (see Fig. 8.5). When the effector cells are depleted, the viral load rebounds to a high level that is similar to the levels observed during acute HIV infection (see Fig. 8.5). The reachability condition (8.9) appropriately reports the failure of the elite control of HIV because the reachability condition becomes positive and has a positive peak corresponding to the peak of the viral load during the depletion period. This is seen in Fig. 8.5. The reachability condition becomes negative temporarily during the depletion period because the limitation in the availability of target cells reduces the viral load for a short time. Consequently, the sliding mode on the manifold $s_E(t) = 0$ and $s_V(t) = 0$ is lost during the depletion period (see Fig. 8.5). The simulation results indicate that the proposed sliding mode control framework appropriately monitors success and failure of elite control of HIV by highlighting a case in which elite control of HIV is sensitive to unmatched biological perturbations and uncertainties.

Next, the robustness of elite control of HIV is investigated in the presence of variations in the number of HIV latently infected cells. In Fig. 8.6 it is seen that when the reachability condition (8.9) is satisfied, elite control of HIV exhibits some robustness to variations in the number of latently infected cells. This suggests that viral clearance is sustained despite variations in the number of latently infected cells when the population of virus-producing infected cells declines.

The robustness of the elite control of HIV is now investigated by considering uncertainty in the activation rate of latently infected cells. Fluctuations in the activation rate of latently infected cells are of interest because the dynamics of the latent reservoir is an important part of HIV infection [3]. The impact of this matched uncertainty is assessed by varying the activation rate of latently infected cells from $w_L = 10^{-4}$ to $w_L = 5 \times 10^{-3}$

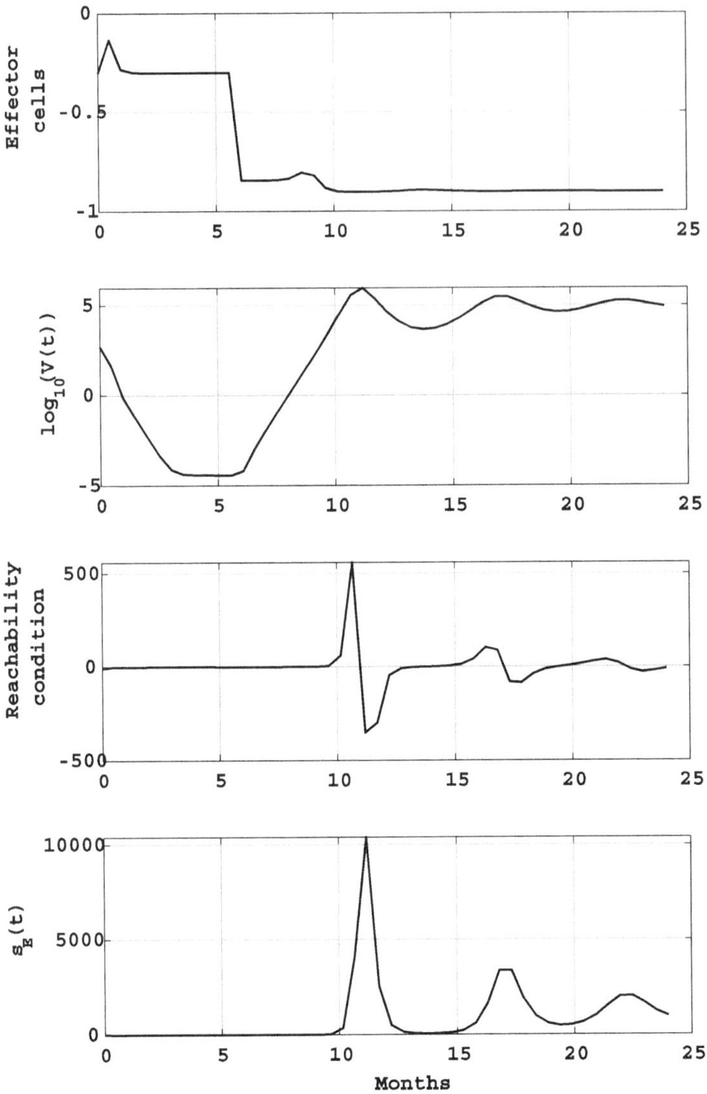

FIGURE 8.5 Effects of the depletion of effector cells on the dynamics of elite control of HIV. Initial conditions: $T(0) = 10^6$; $L(0) = 0$; $I(0) = 10$; $V(0) = 500$; $E(0) = 0.5$. Effector cells are depleted using $-3E$ after 6 months.

day^{-1}. A sliding motion is attained when the reachability condition (8.9) is satisfied (see Fig. 8.7). As the activation rate of latent cells increases, a larger number of latent cells become virus-producing infected cells and this tends to increase the viral load (see Fig. 8.7). When $w_L = 5 \times 10^{-3}$ day^{-1}, the viral load is seen to spike when the reachability condition (8.9)

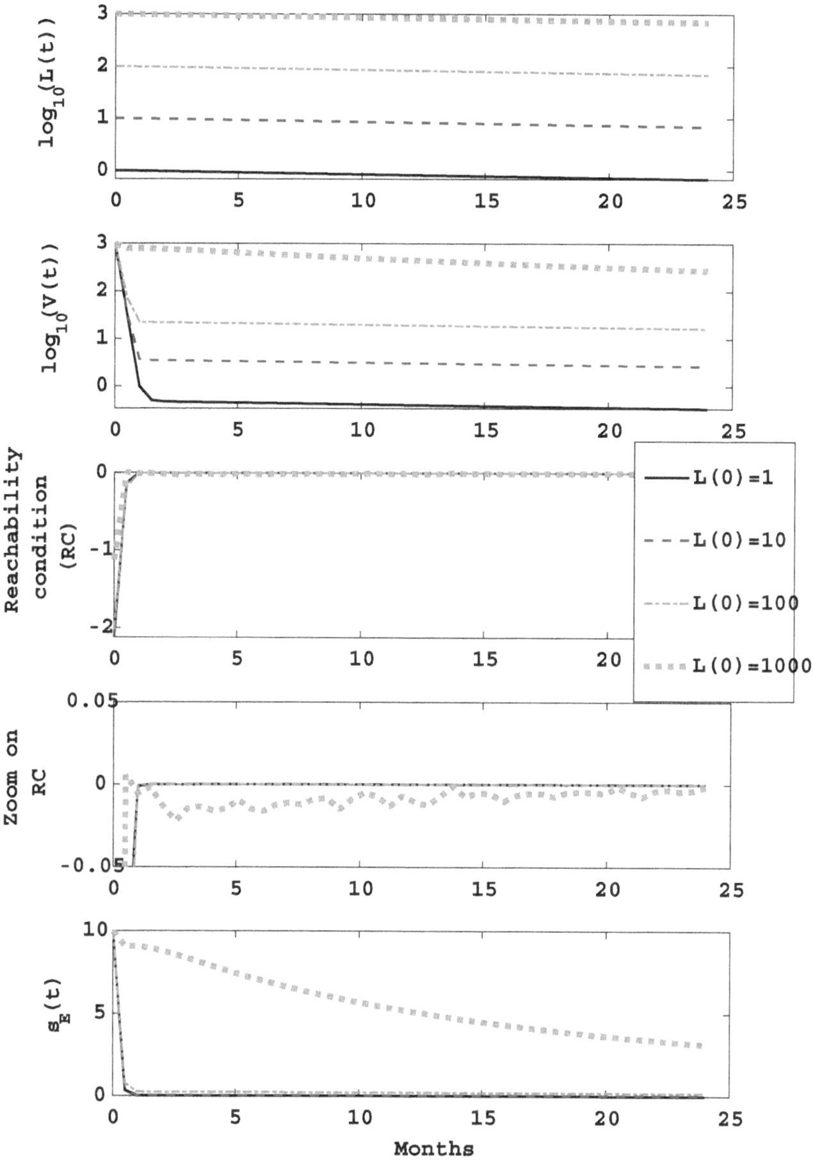

FIGURE 8.6 Robustness to variations in the number of latently infected cells, $L(0)$. Initial conditions: $T(0) = 10^6$; $I(0) = 10$; $V(0) = 10^3$; $E(0) = 0.75$.

is not satisfied. The population of latently infected cells declines due to this high activation rate $w_L = 5 \times 10^{-3}$ day^{-1}. Then the reachability condition (8.9) becomes negative and a sliding mode is enforced on $s_E(t) = 0$ and hence $s_V(t) = 0$. The viral load declines quickly due the rapid drop in

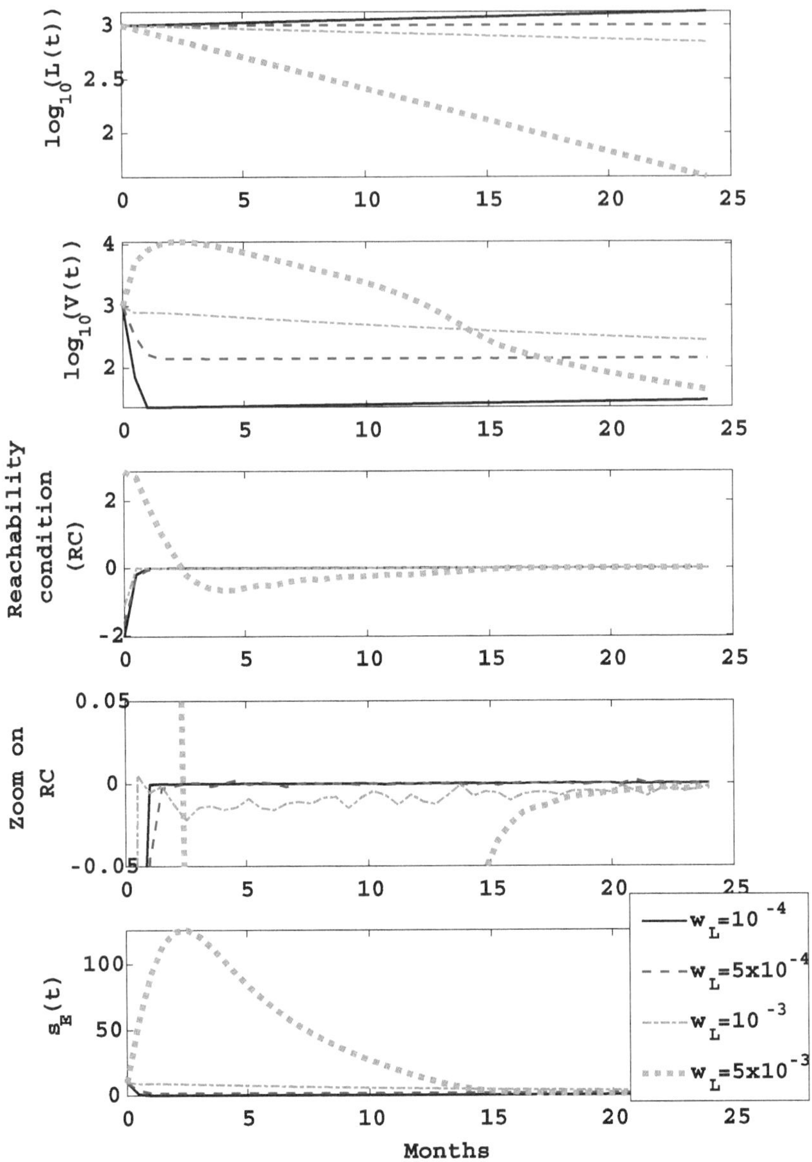

FIGURE 8.7 Robustness to variations in the activation rate of latently infected cells, w_L. $T(0) = 10^6$; $L(0) = 1000$; $I(0) = 10$; $V(0) = 10^3$; $E(0) = 0.75$.

the number of virus-producing infected cells. For $w_L = 5 \times 10^{-3}$ day^{-1}, latent cells generate fewer virus-producing infected cells over time because the number of latent cells declines for this activation rate. These findings suggest that as long as the population of virus-producing infected cells declines, elite control of HIV ensures that the viral load declines in the presence of matched uncertainties such as the activation of latently infected cells.

Collectively, these simulation results validate the analytical analysis and highlight the synergies between the robustness of sliding mode controllers and the robustness of elite control of HIV infection in vivo.

8.6 Conclusion

In this chapter, methods from control theory have been used to analyze the feedback inherent in a model of T cell immunity in the context of HIV infection. The robustness properties of elite control of HIV were investigated by casting this immune control problem within the sliding mode control paradigm. These investigations were conducted on an established model of HIV infection in vivo. The characteristics of elite control of HIV observed in patients have been used to formulate a reachability analysis. The results show that the action of elite control of HIV can be modeled by defining an appropriate sliding mode. The analytical and simulation results demonstrate that elite control of HIV exhibits robustness properties similar to the ones present in a well-designed sliding mode control. The reachability paradigm is useful to classify the impact of perturbations. Indeed, the reachability paradigm predicts and appropriately monitors the success or failure of HIV suppression in the presence of biological perturbations and variations in biological rates. Together, these findings suggest that sliding mode control theory delivers an analytical framework with mature tools that could benefit the study of T cell immunity and treatments against viral infections.

Acknowledgments

Sarah Spurgeon and Anet Anelone gratefully acknowledge the support of the EPSRC via grant reference number EP/J018295/1. Peter Kim and Anet Anelone gratefully acknowledge the support of the Australian Research Council via grant reference number DP180101512 and the University of Sydney – UCL Mobility Scheme.

References

[1] P. Goulder, S.G. Deeks, Hiv control: is getting there the same as staying there?, PLoS Pathogens 14 (11) (2018) 1–18, https://doi.org/10.1371/journal.ppat.1007222.

[2] K. O'Connell, J. Bailey, J. Blankson, Elucidating the elite: mechanisms of control in hiv-1 infection, Trends in Pharmacological Sciences 30 (12) (2009) 631–637, https://doi.org/10.1016/j.tips.2009.09.005, http://www.sciencedirect.com/science/article/pii/S0165614709001539.

[3] A.S. Perelson, R. Ribeiro, Modeling the within-host dynamics of HIV infection, BMC Biology 11 (1) (2013) 96, https://doi.org/10.1186/1741-7007-11-96, http://www.biomedcentral.com/1741-7007/11/96.

[4] O. Lambotte, J. Delfraissy, Hiv controllers: a homogeneous group of hiv-1–infected patients with spontaneous control of viral replication, Pathologie Et Biologie 54 (2007) 566–571, https://doi.org/10.1016/j.patbio.2006.07.035.
[5] J. Blankson, Effector mechanisms in hiv-1 infected elite controllers: highly active immune responses?, in: Twenty-Five Years of Antiretroviral Drug Development: Progress and Prospects, Antiviral Research 85 (1) (2010) 295–302, https://doi.org/10.1016/j.antiviral.2009.08.007, http://www.sciencedirect.com/science/article/pii/S0166354209004525.
[6] A.N. de Recherche sur le Sida (ANRS) PRIMO Study Group, C. Goujard, A. Venet, J. Delfraissy, O. Lambotte, M. Sinet, C. Rouzioux, M. Chaix, C. Deveau, L. Meyer, J. Guergnon, V. Courgnaud, Spontaneous control of viral replication during primary HIV infection: when is "HIV controller" status established?, Clinical Infectious Diseases 49 (6) (2009) 982–986, https://doi.org/10.1086/605504, arXiv: http://oup.prod.sis.lan/cid/article-pdf/49/6/982/962901/49-6-982.pdf.
[7] R.J. De Boer, Which of our modeling predictions are robust?, PLoS Computational Biology 8 (7) (2012) e1002593, https://doi.org/10.1371/journal.pcbi.1002593.
[8] A.S. Perelson, P.W. Nelson, Mathematical analysis of HIV-1 dynamics in vivo, SIAM Review 41 (1998) 3–44.
[9] H. Wu, H. Zhu, H. Miao, A. Perelson, Parameter identifiability and estimation of HIV/AIDS dynamic models, Bulletin of Mathematical Biology 70 (3) (2008) 785–799.
[10] M.A. Stafford, L. Corey, Y. Cao, E.S. Daar, D.D. Hod, A.S. Perelson, Modeling plasma virus concentration during primary HIV infection, Journal of Theoretical Biology 203 (3) (2000) 285–301, https://doi.org/10.1006/jtbi.2000.1076, http://www.sciencedirect.com/science/article/pii/S0022519300910762.
[11] J. Conway, A. Perelson, Post-treatment control of hiv infection, Proceedings of the National Academy of Sciences 112 (17) (2015) 5467–5472, https://doi.org/10.1073/pnas.1419162112, arXiv: http://www.pnas.org/content/112/17/5467.full.pdf, http://www.pnas.org/content/112/17/5467.abstract.
[12] E.A. Hernandez-Vargas, R.H. Middleton, Modeling the three stages in hiv infection, Journal of Theoretical Biology 320 (2013) 33–40, https://doi.org/10.1016/j.jtbi.2012.11.028, https://www.sciencedirect.com/science/article/pii/S0022519312006170.
[13] T. Yang, Impulsive Control Theory, Vol. 272, Springer Science & Business Media, 2001.
[14] G. Conte, C.H. Moog, A.M. Perdon, Algebraic Methods for Nonlinear Control Systems, Springer Science & Business Media, 2007.
[15] C. Edwards, S.K. Spurgeon, Sliding Mode Control: Theory and Applications, CRC Press, 1998.
[16] V.I. Utkin, Sliding Modes in Control Optimization, Springer-Verlag, Berlin, 1992.
[17] E.A. Hernandez-Vargas, P. Colaneri, R.H. Middleton, Switching strategies to mitigate hiv mutation, IEEE Transactions on Control Systems Technology 22 (4) (2014) 1623–1628, https://doi.org/10.1109/TCST.2013.2280920.
[18] A.J.N. Anelone, S. Spurgeon, Prediction of the containment of hiv infection by antiretroviral therapy – a variable structure control approach, IET Systems Biology (August 2016), http://digital-library.theiet.org/content/journals/10.1049/iet-syb.2016.0028.
[19] P.S. Rivadeneira, C.H. Moog, Impulsive control of single-input nonlinear systems with application to hiv dynamics, Applied Mathematics and Computation 218 (2012) 8462–8474.
[20] P.S. Rivadeneira, M. Caicedo, A. Ferramosca, A.H. González, Impulsive zone model predictive control (izmpc) for therapeutic treatments: application to hiv dynamics, in: 2017 IEEE 56th Annual Conference on Decision and Control (CDC), 2017, pp. 4094–4099.
[21] H. Chang, C. Moog, A. Astolfi, P. Rivadeneira, A control systems analysis of hiv prevention model using impulsive input, Biomedical Signal Processing and Control 13 (2014) 123–131, https://doi.org/10.1016/j.bspc.2014.03.008, https://www.sciencedirect.com/science/article/pii/S1746809414000469.

[22] R. Zurakowski, Nonlinear observer output-feedback mpc treatment scheduling for HIV, Biomedical Engineering Online 10 (2011) 40.
[23] R. Zurakowski, A.R. Teel, A model predictive control based scheduling method for HIV therapy, Journal of Theoretical Biology 238 (2) (2006) 368–382, https://doi.org/10.1016/j.jtbi.2005.05.004, http://www.sciencedirect.com/science/article/pii/S0022519305002067.
[24] H.-D. Kwon, Optimal treatment strategies derived from a {HIV} model with drug-resistant mutants, Applied Mathematics and Computation 188 (2) (2007) 1193–1204, https://doi.org/10.1016/j.amc.2006.10.071, http://www.sciencedirect.com/science/article/pii/S0096300306014639.
[25] E.A. Hernandez-Vargas, P. Colaneri, R.H. Middleton, Optimal therapy scheduling for a simplified hiv infection model, Automatica 49 (9) (2013) 2874–2880.
[26] A. Jagarapu, M.J. Piovoso, R. Zurakowski, Optimal control modulation of hiv reservoir formation rate by antigen infusion, in: 2019 IEEE 58th Conference on Decision and Control (CDC), IEEE, 2019, pp. 5662–5667.
[27] A.J.N. Anelone, Y. Orlov, S.K. Spurgeon, Synergies between the dynamics of the immune response of T cells and the variable structure control paradigm, in: Recent Advances in Sliding Modes (RASM), 2015 International Workshop on, 2015, pp. 1–6.
[28] A.J.N. Anelone, S. Spurgeon, Modelling and simulation of the dynamics of the antigen-specific t cell response using variable structure control theory, PLoS ONE 11 (11) (2016) 1–22, https://doi.org/10.1371/journal.pone.0166163.
[29] M. Bofill, G. Janossy, C. Lee, D. MacDonald-Burns, A. Phillips, C. Sabin, A. Timms, M. Johnson, P. Kernoff, Laboratory control values for cd4 and cd8 t lymphocytes. Implications for hiv-1 diagnosis, Clinical and Experimental Immunology 88 (2) (1992) 243–252.
[30] H. Mohri, S. Bonhoeffer, S. Monard, A.S. Perelson, D.D. Ho, Rapid turnover of T lymphocytes in SIV-infected rhesus macaques, Science 279 (5354) (1998) 1223–1227.
[31] J.M. Conway, D. Coombs, A stochastic model of latently infected cell reactivation and viral blip generation in treated hiv patients, PLoS Computational Biology 7 (4) (2011) e1002033.
[32] L. Rong, A.S. Perelson, Asymmetric division of activated latently infected cells may explain the decay kinetics of the HIV-1 latent reservoir and intermittent viral blips, Mathematical Biosciences 217 (1) (2009) 77–87.
[33] M. Markowitz, M. Louie, A. Hurley, E. Sun, M. Di Mascio, A.S. Perelson, D.D. Ho, A novel antiviral intervention results in more accurate assessment of human immunodeficiency virus type 1 replication dynamics and T-cell decay in vivo, Journal of Virology 77 (8) (2003) 5037–5038.
[34] B. Ramratnam, S. Bonhoeffer, J. Binley, A. Hurley, L. Zhang, J.E. Mittler, M. Markowitz, J.P. Moore, A.S. Perelson, D.D. Ho, Rapid production and clearance of HIV-1 and hepatitis C virus assessed by large volume plasma apheresis, The Lancet 354 (9192) (1999) 1782–1785.
[35] M.P. Davenport, R.M. Ribeiro, A.S. Perelson, Kinetics of virus-specific cd8+ T cells and the control of human immunodeficiency virus infection, Journal of Virology 78 (18) (2004) 10096–10103.
[36] S. Bonhoeffer, M. Rembiszewski, G.M. Ortiz, D.F. Nixon, Risks and benefits of structured antiretroviral drug therapy interruptions in HIV-1 infection, AIDS 14 (15) (2000) 2313–2322.
[37] P.L. Johnson, B.F. Kochin, M.S. McAfee, I.M. Stromnes, R.R. Regoes, R. Ahmed, J.N. Blattman, R. Antia, Vaccination alters the balance between protective immunity, exhaustion, escape, and death in chronic infections, Journal of Virology 85 (11) (2011) 5565–5570.
[38] N.R. Klatt, N. Chomont, D.C. Douek, S.G. Deeks, Immune activation and HIV persistence: implications for curative approaches to HIV infection, Immunological Reviews 254 (1) (2013) 326–342, https://doi.org/10.1111/imr.12065, arXiv: https://onlinelibrary.wiley.com/doi/pdf/10.1111/imr.12065, https://onlinelibrary.wiley.com/doi/abs/10.1111/imr.12065.

CHAPTER

9

A stochastic model for hepatitis C viral infection dynamics with the innate immune response

Mauricio Castaño-Arcila[b] *and Jesús Rodríguez-González*[a]

[a]Center for Research and Advanced Studies of the National Polytechnic Institute, Monterrey Unit, Parque PIIT, Apodaca, NL, Mexico, [b]Francisco José de Caldas District University, Bogotá, Colombia

9.1 Introduction

Hepatitis C virus (HCV) is a positive strand RNA molecule of 9.6 kbp length belonging to the Flaviviridae family. Naturally HCV infects only humans, but chimpanzees can be experimentally infected [1]. HCV infection is asymptomatic, and in 70% of infected individuals, the virus persists and the infection progresses to chronic liver diseases. To guard against viral infection the host cell has multiple restriction strategies to limit the viral infection and many of these strategies are associated with induction of the interferon (IFN) [2]. Antiviral signaling begins when IFNs bind to transmembrane receptors and activate the JAK-STAT signaling pathway to regulate the transcription of interferon-stimulated genes (ISGs) in the nucleus [3]. More than 300 ISGs are regulated by IFN through the JAK-STAT signaling pathway, and 93 ISG proteins mediate the anti-HCV effect [4]. Metz and colleagues identified seven ISG proteins as suppressors of the HCV replication [5]. However, for most ISG products, very little is known about their targets and their mode of action. Several mechanisms of viral interference with the hepatic IFN system have been studied [2]. HCV could inhibit IFN-induced signal transduction through the JAK-STAT pathway, lowering the translation of ISG mRNAs [3,6]. A graphical representation

Feedback Control for Personalized Medicine
https://doi.org/10.1016/B978-0-32-390171-0.00018-4

of the interaction between HCV and the innate immune response is given in Fig. 9.1.

HCV intracellular reactions are strongly affected by the randomness of molecular events. This randomness arises from the beginning of the infection by the discrete number of infecting viral particles reaching the cell [7]. Similarly, a low copy number of ISG mRNAs are present in the cell at basal state [8]. Ivanisenko and colleagues consider noise effects in HCV replication of drug-resistant mutants [9]. The dynamics of the viral RNA strands and the innate immune response under noisy environments are not yet fully understood.

Numerous theoretical investigations to study several aspects of HCV epidemiology have been reported. Some mathematical models are focused on capturing the dynamics of infection contrasting with patient data from varying doses of IFN [10,11]. A mathematical model that considers intracellular HCV replication was developed by Dahari and colleagues [12]. This model describes viral RNA replication in the absence of inhibitors establishing the steady state after transfection. Extended models were suggested to explain the difference in HCV permissiveness in different cell lines [13] or the assembly of new viral particles in a single infected cell [14], or focused on the HCV full life cycle, including translation, replication, and assembly [15]. A model that considers the roles of ISG proteins in inhibiting viral translation is reported by Padmanabhan and colleagues [16]. These models describe key elements of the viral infection and have improved our quantitative understanding of the infection. However, the interplay between the viral replication cycle and IFN activity has received little attention. This work is devoted to partially filling this gap in knowledge. In particular, we address two scientific questions. (i) Can the interactions between HCV and the innate immune response exhibit bimodal behaviors? (ii) Furthermore, how can IFITM1 overexpression modulate the interplay between the virus and the innate immune response?

9.2 Biological background

Mathematical modeling has been applied to biological systems for decades. Gene regulatory models generally employ analytical approaches. For example, the models can include terms relating to the binding of transcription factors and RNA polymerase to the DNA, cooperative and inhibitory interactions between transcription factors, mRNA and protein degradation, and mRNA translation rates [17,18]. Still, concerning gene expression, modeling gene regulation is central to such efforts because gene expression is at the nexus of many biological processes, and subtle changes in regulatory protein levels or links can underlie human diseases [19].

On the other hand, a virus is a submicroscopic infectious agent that replicates only inside the living cells of an organism. When infected, a host cell is forced to rapidly produce thousands of copies of the original virus. In the process of infecting a cell, viruses exist in the form of independent molecules (virions) consisting of the genetic material: DNA or RNA molecules that encode the structure of the proteins by which the virus acts [20]. During the last two decades, there has been broad interest in mRNA vaccines for personalized cancer treatments and vaccines for infectious diseases [21].

9.2.1 Gene regulation models

All known living systems are made up of one basic structural and functional unit: the cell. It is a miniature, self-reproducing biochemical machine. Organisms may consist of just one cell, or they may be multicellular. Thus, one may view cell life as a collection of biochemical networks of interactions among molecules: DNA, RNA, and proteins. These networks process environmental signals, induce appropriate cellular responses and internal sequence events such as gene expression, and activate signaling pathways allowing cells and entire organisms to perform their essential functions.

The central dogma of molecular biology describes the flow of information in the cell. Frances Crick first described the central dogma in 1958 to explain how information flows from the language of nucleic acids to the language of proteins [22]. Cells copy their DNA by a process called DNA replication. DNA replication occurs whenever a cell is going to reproduce itself by either mitosis or binary fission. Inside a cell, the information for biomolecules is encoded in the genetic material called nucleic acids: DNA (deoxyribonucleic acid) and RNA (ribonucleic acid) [23].

Transcription process. The information in DNA is copied by transcription to make mRNA. The single strand of RNA nucleotides is generated by complementarity with one strand of a DNA molecule. During transcription, the double helix of DNA is opened, and one strand of DNA is used as a pattern for the construction of an RNA molecule. Several types of RNA molecules have different functions in the cell. The RNA molecules either perform a function directly for the cell or copy important cell information [24].

Translation process. mRNA carries the information for proteins to the ribosome. One mRNA molecule and many tRNA molecules work together to translate the mRNA message and build proteins. Translation occurs in three phases. (a) Initiation of translation begins when the small subunit of the ribosome binds to the ribosomal binding site on the mRNA. Then, the aminoacyl tRNA that carries methionine binds by its anticodon to the start codon. Finally, the large subunit of the ribosome binds to the small subunit and mRNA. (b) Elongation of translation occurs as amino acids are brought together and joined to form a polypeptide chain. Elongation has several steps, which repeat until the polypeptide chain is complete. (c) Termination of translation begins when a stop codon enters the A site of the ribosome. Stop codons are not recognized by aminoacyl tRNAs. Instead, the stop codon is recognized by an enzyme called release factor. Release factor enters the ribosome and catalyzes the breaking of the bond between the growing polypeptide chain and the tRNA that holds it. Release factor releases the polypeptide chain from the ribosome to be converted into a finished protein [24].

Gene regulation is the process of cells choosing which genes to use and which not to use. Most types of cells have a complete set of genes, yet cells use only a fraction of those genes at any given moment. Cells may change what genes they are using depending on signals or changes in the environment. Constitutive genes are genes that are expressed all the time. Regulated genes are turned on and off as they are needed by the cell. Gene regulation at the level of transcription, called transcriptional control, occurs in both bacteria and eukaryotes. A cell can use several mechanisms to block the translation of an mRNA. The readers interested in learning more about the importance of positive and negative feedback regulatory loops in the dynamic regulation of signaling pathways are referred to two excellent books [25,26].

9.2.2 mRNA vaccines for infectious diseases

Our bodies are continuously exposed to molecules that may indicate disease or parasitic invasion. The immune system is responsible for detecting and clearing such molecules and controlling the underlying causes. Thus, manipulating the immune system is an important component of many therapeutic applications against infectious and degenerative diseases. Vaccines work by training the body to recognize and respond to the proteins produced by disease-causing organisms, such as a virus or bacteria. Traditional vaccines are made up of small or inactivated doses of the whole disease-causing organism or the proteins that it produces, which are introduced into the body to provoke a response from the immune system [27].

There are two parts to our immune system: innate (the defenses we are born with) and acquired (which we develop as we contact pathogens). Classical vaccines usually work with the acquired immune system, and the innate immune system is activated by another ingredient, called an adjuvant. Interestingly, mRNA in vaccines could also trigger the innate immune system, providing an extra layer of defense without adding adjuvants [28].

Gene therapy is the transfer of genetic material to a patient to treat a disease. mRNA vaccines trick the body into producing some of the viral proteins themselves. They work by using mRNA, which is the molecule that essentially puts DNA instructions into action. Inside a cell, mRNA is used as a template to build a protein [29]. mRNA vaccines are a product of decades of study on RNA therapies and treatment by medical scientists. mRNA therapies are being used to develop personalized cancer treatments and vaccines for infectious diseases such as SARS-CoV-2, Dengue, and Zika virus. Researchers are also exploring whether mRNA treatments can be used as protein replacement therapies for rare conditions such as the blood-clotting disorder hemophilia. The readers interested in learning more about gene therapy are referred to the excellent review papers [30–32].

9.2.3 Stochastic process

At macroscopic scales, the processes of life appear highly deterministic. There are two dominant paradigms used in mathematical modeling of biochemical reaction networks: the deterministic approach, using numerical simulations of nonlinear ordinary differential equations (including mass-action-type, power-law, or Michaelis–Menten models) [33], and the stochastic approach based on a master equation and stochastic simulations [34]. A mathematical model is called deterministic if its behavior is exactly reproducible. Although the behavior of a deterministic model is dependent on a specified set of conditions, no other factors have any influence, so that repeated simulations under the same conditions are always in perfect agreement. For example, modeling reactions in biochemical networks uses the law of mass action which results in a set of differential equations that describe the evolution of concentrations of species adopted by the network over time.

At mesoscopic scales, in the case of biological processes involving molecular species at very low copy numbers, a potentially significant stochastic effect may arise due to the random variations in the numbers of molecules present in specific molecular processes. However, this stochasticity or noise is neglected in deterministic chemical rate equations represented by differential equation models because such effects are averaged over large numbers of molecules in chemical reactions. An example of a noise system is the process of transcription [35]. Usually, one or two copies of a particular DNA regulatory site may be present in each cell. A second example is the viral load dynamics for infections; they initiate with low virus copy. Those cellular events are discrete and random; they are supported by many recent experiments that have revealed cell–cell variations in isogenic cell populations of transcription and translation [36–38].

Thus, in stochastic formulation of chemical kinetics, molecular species are characterized by their probability density function, which quantifies the amount of fluctuations around a specific mean value. Unfortunately, the theoretical machinery required to analyze stochastic molecular networks rigorously is much less well developed than in deterministic approaches. The reactant species dynamics result in a behavior that is discrete and stochastic rather than continuous and deterministic; the dynamic is driven by the chemical master equation (CME). It represents the time evolution for the probability function of the system state. Nevertheless, the CME cannot be analytically or numerically solved. However, we can generate numerical trajectories in state space (realizations) of the stochastic process defined by the CME by using the stochastic simulation algorithm (SSA). The groundbreaking work of Gillespie presented an algorithm for numerically generating sample trajectories of the abundances of chemical species in chemical reaction networks. The SSA can easily be implemented in any programming or scripting language that has a pseudorandom number generator [39,40].

The SSA involves the following steps: (1) Initialize the system at $t = 0$ with initial numbers of molecules for each chemical species. (2) Generate the time until the following reaction occurs. (3) Determine which reaction occurs at that time. (4) Update the time and state to reflect the previous two choices [41]. The readers interested in learning more about stochastic approaches for systems biology are referred to the excellent book [34].

9.3 Stochastic model

A stochastic mathematical model of the molecular interaction network (Fig. 9.1) has been developed following Gillespie's derivation of the minimal mathematical model reported in [17]. We considered the set of reactions shows in Table 9.1, which involves the production and loss of the positive and negative strand HCV RNA, the ISG mRNAs, and the ISG proteins.

The model considers the time evolution of the positive strand HCV RNA, V_p, and the negative strand HCV RNA, V_N. Furthermore, the innate immune response is considered by the transcription and translation processes of the following seven ISGs: the *IFIT3* mRNA, M_1, the *IFITM1* mRNA, M_2, the *IFITM3* mRNA, M_3, the *TRIM14* mRNA, M_4, the *Viperin* mRNA, M_5, the *PLSCR1* mRNA, M_6, the *RNaseL* mRNA, M_7, the IFITM3 protein, P_1, the IFITM1 protein, P_2, the IFITM3 protein, P_3, the TRIM14 protein, P_4, the Viperin protein, P_5, the PLSCR1 protein, P_6, and the RNaseL protein, P_7. The SSA was implemented using the direct method by Gillespie [40].

All parameter values used have biological relevance and were obtained from published experimental data. mRNA transcription rates were estimated using the ISG mRNA sequence lengths from the NCBI database and the rate of stable RNA synthesis (306,000 nucleotides per hour) [42]. Protein translation rates were obtained using the ISG protein length from the NCBI database and the global translation rate (39,600 amino acids per

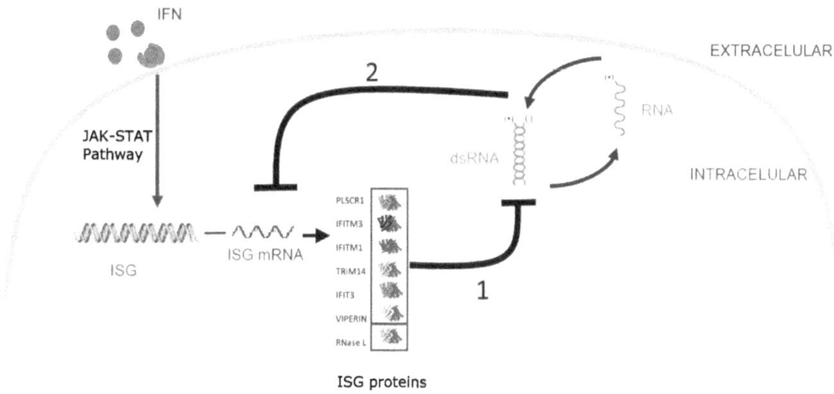

FIGURE 9.1 Graphical representation of the interactions between the innate immune response and HCV in an infected cell. IFN is produced in response to viral infection. The HCV replication process produces positive and negative RNA strands. Following exposure to IFN, the JAK-STAT signaling pathway triggers the expression of ISGs. The translation of ISG mRNA yields ISG proteins that control HCV replication by inhibiting the HCV life cycle (negative feedback 1). HCV proteins form a negative feedback loop that inhibits IFN response (negative feedback 2).

hour) reported by [43]. Table 9.2 shows the parameter values used in this study.

9.4 Results

Recall that at the beginning of the HCV infection, there is a low copy number for both the innate immune response components and HCV biomolecules [7,46]. As the concentration of these molecules is low, stochasticity may play an important role in the final viral outcome. We numerically calculated the steady-state histograms of our stochastic model to address whether the molecular machinery involved in a cell infected with HCV and the innate immune response generate bimodal dynamics. The system was solved using Gillespie's algorithm [40], and we carried out 1000 independent realizations. The stochastic model was numerically solved for a simulation time of 200 h in order to allow us to compare with deterministic results reported in [16] [17]. As initial conditions, we set IFN = 100 IU/ml and $MOI = 3$. Fig. 9.2 illustrates that the molecular system reproduces bimodal dynamics. Two modes that correspond to the clearance and persistent states of the molecular model can be observed in the histogram. The coefficient of variation is a standardized measure of the dispersion of a probability distribution, which considers the median and standard deviation of the data [35]. The first mode has a median of

TABLE 9.1 System reactions for HCV-infected hepatocytes.

Reaction	Reaction rate	Reaction definition
$V_N \to V_P$	$k_1(\prod_m \frac{K_{rm}}{K_{rm}+[P_m]})(1-\frac{[V_P]+[V_N]}{Kc})[V_N]$	V_P synthesis from V_N
$V_P \to \phi$	$\frac{\mu_1}{\frac{K_{d_7}}{K_{d_7}+[P_7]}}[V_P]$	V_P degradation rate
$V_P \to V_N$	$k_2(\prod_m \frac{K_{rm}}{K_{rm}+[P_m]})(1-\frac{[V_P]+[V_N]}{Kc})[V_P]$	V_N synthesis from V_P
$V_N \to \phi$	$\mu_2[V_N]$	V_N degradation rate
$\phi \to m_1$	$k_{m_1}\frac{[IFN]}{K_{i_{IFIT3}}+[IFN]}\frac{K_H}{K_H+[V_P]}$	*IFIT3* transcription rate
$\phi \to m_2$	$k_{m_2}\frac{[IFN]}{K_{i_{IFITM1}}+[IFN]}\frac{K_H}{K_H+[V_P]}$	*IFITM1* transcription rate
$\phi \to m_3$	$k_{m_3}\frac{[IFN]}{K_{i_{IFITM3}}+[IFN]}\frac{K_H}{K_H+[V_P]}$	*IFITM3* transcription rate
$\phi \to m_4$	$k_{m_4}\frac{[IFN]}{K_{i_{TRIM14}}+[IFN]}\frac{K_H}{K_H+[V_P]}$	*TRIM14* transcription rate
$\phi \to m_5$	$k_{m_5}\frac{[IFN]}{K_{i_{Viperin}}+[IFN]}\frac{K_H}{K_H+[V_P]}$	*Viperin* transcription rate
$\phi \to m_6$	$k_{m_6}\frac{[IFN]}{K_{i_{PLSCR1}}+[IFN]}\frac{K_H}{K_H+[V_P]}$	*PLSCR1* transcription rate
$\phi \to m_7$	$k_{m_7}\frac{[IFN]}{K_{i_{RNaseL}}+[IFN]}\frac{K_H}{K_H+[V_P]}$	*RNaseL* transcription rate
$M_1 \to \phi$	$\mu_{m_1}[M_1]$	*IFIT3* mRNA degradation rate
$M_2 \to \phi$	$\mu_{m_2}[M_2]$	*IFITM1* mRNA degradation rate
$M_3 \to \phi$	$\mu_{m_3}[M_3]$	*IFITM3* mRNA degradation rate
$M_4 \to \phi$	$\mu_{m_4}[M_4]$	*TRIM14* mRNA degradation rate
$M_5 \to \phi$	$\mu_{m_5}[M_5]$	*Viperin* mRNA degradation rate
$M_6 \to \phi$	$\mu_{m_6}[M_6]$	*PLSCR1* mRNA degradation rate
$M_7 \to \phi$	$\mu_{m_7}[M_7]$	*RNaseL* mRNA degradation rate
$M_1 \to P_1$	$k_{p_1}[M_1]$	IFIT3 translation rate
$M_2 \to P_2$	$k_{p_2}[M_2]$	IFITM1 translation rate
$M_3 \to P_3$	$k_{p_3}[M_3]$	IFITM3 translation rate
$M_4 \to P_4$	$k_{p_4}[M_4]$	TRIM14 translation rate
$M_5 \to P_5$	$k_{p_5}[M_5]$	Viperin translation rate
$M_6 \to P_6$	$k_{p_6}[M_6]$	PLSCR1 translation rate
$M_7 \to P_7$	$k_{p_7}[M_7]$	RNaseL translation rate
$P_1 \to \phi$	$\mu_{p_1}[P_1]$	IFIT3 protein degradation rate
$P_2 \to \phi$	$\mu_{p_2}[P_2]$	IFITM1 protein degradation rate
$P_3 \to \phi$	$\mu_{p_3}[P_3]$	IFITM3 protein degradation rate
$P_4 \to \phi$	$\mu_{p_4}[P_4]$	TRIM14 protein degradation rate
$P_5 \to \phi$	$\mu_{p_5}[P_5]$	Viperin protein degradation rate
$P_6 \to \phi$	$\mu_{p_6}[P_6]$	PLSCR1 protein degradation rate
$P_7 \to \phi$	$\mu_{p_7}[P_7]$	RNaseL protein degradation rate

TABLE 9.2 Model parameter values.

Parameter	Value	Unit	Reference
k_1	9.46×10^{-1}	h^{-1}	[16]
k_2	4.73×10^{-2}	h^{-1}	[16]
Kc	317.54	molecules $cell^{-1}$	[16]
μ_1	4.62×10^{-2}	h^{-1}	[44]
μ_2	5.77×10^{-2}	h^{-1}	[44]
k_{m_1}	124.03	molecules $cell^{-1}$ h^{-1}	RefSeq Accession NM_001031683.3
k_{m_2}	417.46	molecules $cell^{-1}$ h^{-1}	RefSeq Accession NM_003641.3
k_{m_3}	451.33	molecules $cell^{-1}$ h^{-1}	RefSeq Accession NM_021034.2
k_{m_4}	171.05	molecules $cell^{-1}$ h^{-1}	RefSeq Accession NM_033220.1
k_{m_5}	107.26	molecules $cell^{-1}$ h^{-1}	RefSeq Accession AF442151.1
k_{m_6}	137.34	molecules $cell^{-1}$ h^{-1}	RefSeq Accession NM_021105.2
k_{m_7}	71.86	molecules $cell^{-1}$ h^{-1}	RefSeq Accession NM_021133.3
Ki_{IFIT3}	7.12	IU ml^{-1}	[16]
Ki_{IFITM1}	45.09	IU ml^{-1}	[16]
Ki_{IFITM3}	109.77	IU ml^{-1}	[16]
Ki_{TRIM14}	20.66	IU ml^{-1}	[16]
$Ki_{Viperin}$	139.86	IU ml^{-1}	[16]
Ki_{PLSCR1}	14.31	IU ml^{-1}	[16]
Ki_{RNaseL}	20.65	IU ml^{-1}	[16]
μ_{p_1}	2.9×10^{-2}	h^{-1}	[45]
μ_{p_2}	6.19×10^{-2}	h^{-1}	[45]
μ_{p_3}	6.19×10^{-2}	h^{-1}	[45]
μ_{p_4}	3.75×10^{-2}	h^{-1}	[45]
μ_{p_5}	6.08×10^{-2}	h^{-1}	[45]
μ_{p_6}	6.08×10^{-2}	h^{-1}	[45]
μ_{p_7}	2.39×10^{-2}	h^{-1}	[45]
μ_{m_1}	9.27×10^{-2}	h^{-1}	[45]
μ_{m_2}	6.21×10^{-2}	h^{-1}	[45]
μ_{m_3}	7.47×10^{-2}	h^{-1}	[45]
μ_{m_4}	1.59×10^{-1}	h^{-1}	[45]
μ_{m_5}	3.15×10^{-2}	h^{-1}	[45]
μ_{m_6}	1.06×10^{-1}	h^{-1}	[45]
μ_{m_7}	6.75×10^{-2}	h^{-1}	[45]
Kr_{IFIT3}	1.9×10^{5}	molecules $cell^{-1}$	[16]
Kr_{IFITM1}	1.9×10^{5}	molecules $cell^{-1}$	[16]
Kr_{IFITM3}	1.12×10^{6}	molecules $cell^{-1}$	[16]
Kr_{TRIM14}	2.64×10^{6}	molecules $cell^{-1}$	[16]
$Kr_{Viperin}$	1.33×10^{6}	molecules $cell^{-1}$	[16]
Kr_{PLSCR1}	1.16×10^{6}	molecules $cell^{-1}$	[16]
K_{d_7}	4.67×10^{7}	molecules $cell^{-1}$	[16]
K_H	0.5	molecules $cell^{-1}$	[17]
k_{p_1}	80.81	molecules $cell^{-1}$ h^{-1}	RefSeq Accession NP_001540.2

continued on next page

TABLE 9.2 (*continued*)

Parameter	Value	Unit	Reference
k_{p_2}	316.80	molecules cell^{-1} h^{-1}	RefSeq Accession NP_003632.3
k_{p_3}	297.74	molecules cell^{-1} h^{-1}	RefSeq Accession AFF60355.1
k_{p_4}	89.59	molecules cell^{-1} h^{-1}	RefSeq Accession NP_150088.1
k_{p_5}	109.69	molecules cell^{-1} h^{-1}	RefSeq Accession AAL50053.1
k_{p_6}	124.57	molecules cell^{-1} h^{-1}	RefSeq Accession NP_066928.1
k_{p_7}	60.73	molecules cell^{-1} h^{-1}	RefSeq Accession AAH90934.1
IFN	0–200	IU ml^{-1}	Varied in this study

3.10 molecules per cell and a coefficient of variation of 2.57, representing the clearance state. Furthermore, the persistence state has a media of 153.4 molecules per cell with a coefficient of variation of 0.18. These values are consistent with results obtained in [16] [17].

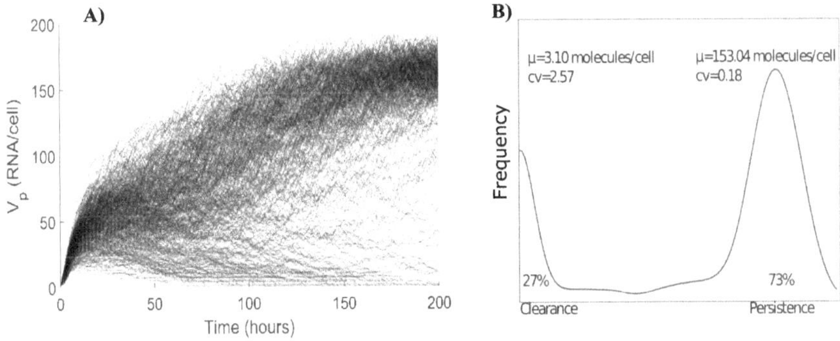

FIGURE 9.2 Bimodal distribution of the virus load. (A) Viral infection dynamics realizations were performed for 1000 individual cells. (B) Clearance and persistent states are present in the histogram at 200 h. IFN dose of 100 IU/ml.

The use of mRNA as a therapeutic molecule for gene therapy purposes has grown recently [21,47,29,48]. To analyze the effect of the overexpression of ISG genes in cells with HCV infection, we simulated the effect of these therapies in our model. We simulated one in silico cell line that describes the overexpression for *IFITM1* mRNA. For the tested simulated cells, ISG overexpression was modeled by replacing the *IFITM1* mRNA propensity function by

$$\gamma_i k_{m_2} \frac{[IFN]}{K_{i_{IFITM1}} + [IFN]} \frac{K_H}{K_H + [V_P]}. \tag{9.1}$$

The overexpression factor is a free parameter calculated with the deterministic model to obtain the clearance state at $IFN \geq 100$ IU/ml [49]. The calculated value is $\gamma_i = 1.8$. It corresponds to the nominal value, which

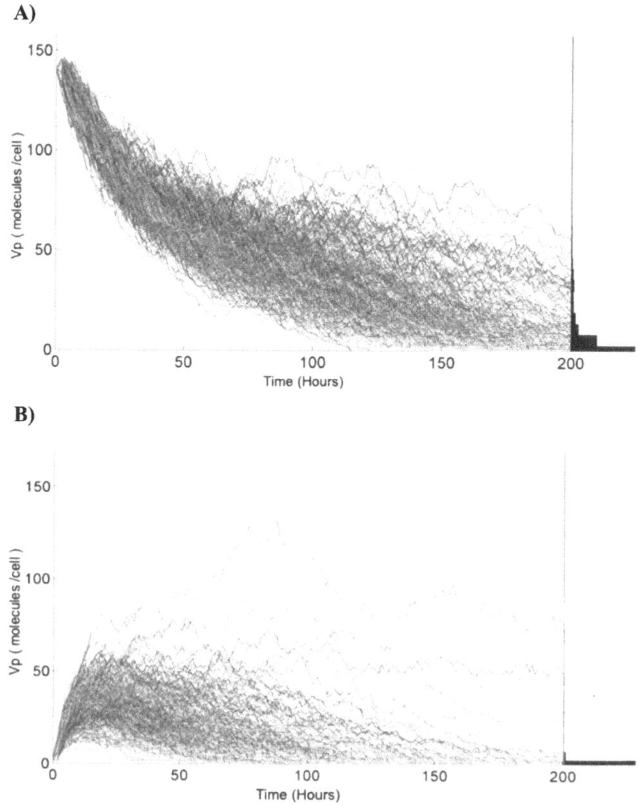

FIGURE 9.3 Overexpressing mRNA in cells. (A) Effect of overexpressing *IFITM1* mRNA in cells with high HCV load. (B) Effect of overexpressing *IFITM1* mRNA in cells with low HCV load.

produces a unimodal distribution and represents 80% of protein overexpression. We next conducted a comparative study. Viral dynamics of a cell line with and without a high viral load in the presence of IFITM1 overexpression was modeled. We numerically solved the stochastic model with the same initial conditions reported above. Fig. 9.3A shows the simulations for hepatocytes infected with HCV using an initial viral load of 140 molecules per cell. Otherwise, the system responds using an initial low viral load concentration of HCV negative and positive strands, three molecules per cell, respectively (see Fig. 9.3B). IFITM1 overexpression leads the system to a clearance state. In the histograms, a unimodal distribution can be observed for low and high initial viral loads. Particularly in this work, IFITM1 protein overexpression is sufficient to strengthen the ISG protein's negative feedback over the HCV and force the cell's clearance state.

9.5 Conclusion

This study developed a mathematical model that considers the molecular machinery involved in the interaction between HCV and the innate immune response. We focused on the feedback mechanisms between HCV and the innate immune response.

Our first task was to investigate the dynamics of the viral positive strand RNA in infected cells. Fig. 9.2 shows the results for hepatocytes infected with HCV. These findings are consistent with experimental reports that show suppression of ISG protein induction in HCV-infected cells [49]. Additionally, we determined that the positive strand RNA virus peak is obtained at 2–4 days postinfection. Stochastic simulations were implemented, and the positive strand viral RNA concentration was tracked until 200 hours postinfection. This long simulation time was used to ensure stability and cover the HCV replication cycle. Our simulation results show that HCV induces a bimodal distribution of the positive strand viral RNA concentration. Bimodality has been theoretically studied in HIV [50,51], and previous reports have proved that bistability is neither sufficient nor necessary for bimodal distributions in a gene regulatory population [52]. Reports for HIV have shown that a deterministic model approximates the monostable population average of a large population of GFP molecules, and the stochastic model presents bistability [53].

Finally, we investigated the use of mRNA overexpression for a specific ISG in combination with IFN therapy against HCV. According to our results, IFN as a single agent is effective against HCV at high concentrations. However, it is known that high levels of IFN can produce a cytokine storm [54]. In contrast, gene therapy in the form of mRNA to overexpress ISGs was effective against HCV (see Fig. 9.3). Then, IFITM1 was the therapeutic target. Our simulation predictions can be compared to experimental reports [16,55,56].

Our analysis predicts bimodality in a specific IFN range. We observed that an intervention strategy involving IFN therapy and gene therapy consisting of overexpressing the IFITM1 protein predicts the virus clearance outcome.

Conflict of interest statement

The authors declare that the research was conducted in the absence of any commercial or financial relationships that could be construed as a potential conflict of interest.

Author contributions

MCA and JRG conceived and designed the research. MCA performed the simulations. MCA performed the formal analysis. MCA and JRG wrote the manuscript. JRG supervised the research and revised the manuscript critically. All authors read and approved the final manuscript.

Funding

This research was partially supported by Consejo Nacional de Ciencia y Tecnología (CONACYT, Mexico) with fellowships to MCA.

References

[1] R.E. Lanford, C. Bigger, S. Bassett, G. Klimpel, The chimpanzee model of hepatitis C virus infections, ILAR Journal 42 (2) (2001) 117–126, https://doi.org/10.1093/ilar.42.2.117.

[2] M.H. Heim, 25 years of interferon-based treatment of chronic hepatitis C: an epoch coming to an end, Nature Reviews. Immunology 13 (1) (2013) 535–542, https://doi.org/10.1038/nri3463.

[3] N.J. Stevenson, N.M. Bourke, E.J. Ryan, M. Binder, L. Fanning, J.A. Johnston, J.E. Hegarty, A. Long, C. O'Farrelly, Hepatitis C virus targets the interferon-α JAK/STAT pathway by promoting proteasomal degradation in immune cells and hepatocytes, FEBS Letters 587 (10) (2013) 1571–1578, https://doi.org/10.1016/j.febslet.2013.03.041.

[4] H. Zhao, W. Lin, K. Kumthip, D. Cheng, D.N. Fusco, O. Hofmann, N. Jilg, A.W. Tai, K. Goto, L. Zhang, W. Hide, J.Y. Jang, L.F. Peng, R.T. Chung, A functional genomic screen reveals novel host genes that mediate interferon-alpha's effects against hepatitis C virus, Journal of Hepatology 56 (2) (2012) 326–333, https://doi.org/10.1016/j.jhep.2011.07.026.

[5] P. Metz, A. Reuter, S. Bender, R. Bartenschlager, Interferon-stimulated genes and their role in controlling hepatitis C virus, Journal of Hepatology 59 (6) (2013) 1331–1341, https://doi.org/10.1016/j.jhep.2013.07.033.

[6] L.-J. Zhao, S.-F. He, Y. Liu, P. Zhao, Z.-Q. Bian, Z.-T. Qi, Inhibition of STAT pathway impairs anti-hepatitis C virus effect of interferon alpha, Cellular Physiology and Biochemistry 40 (1–2) (2016) 77–90, https://doi.org/10.1159/000452526.

[7] B. Schmid, M. Rinas, A. Ruggieri, E.G. Acosta, M. Bartenschlager, A. Reuter, W. Fischl, N. Harder, J.P. Bergeest, M. Flossdorf, K. Rohr, T. Höfer, R. Bartenschlager, Live cell analysis and mathematical modeling identify determinants of attenuation of Dengue virus 2'-O-methylation mutant, PLoS Pathogens 11 (12) (2015) 1–36, https://doi.org/10.1371/journal.ppat.1005345.

[8] W.M. Schneider, M.D. Chevillotte, C.M. Rice, Interferon-stimulated genes: a complex web of host defenses, Annual Review of Immunology 32 (2014) 513–545, https://doi.org/10.1146/annurev-immunol-032713-120231.

[9] N.V. Ivanisenko, E.L. Mishchenko, I.R. Akberdin, P.S. Demenkov, V.A. Likhoshvai, K.N. Kozlov, D.I. Todorov, V.V. Gursky, M.G. Samsonova, A.M. Samsonov, D. Clausznitzer, L. Kaderali, N.A. Kolchanov, V.A. Ivanisenko, A new stochastic model for subgenomic hepatitis C virus replication considers drug resistant mutants, PLoS ONE 9 (3) (2014) 1–17, https://doi.org/10.1371/journal.pone.0091502.

[10] A.U. Neumann, N.P. Lam, H. Dahari, D.R. Gretch, T.E. Wiley, T.J. Layden, A.S. Perelson, Hepatitis C viral dynamics in vivo and the antiviral efficacy of interferon-α therapy, Science 282 (5386) (1998) 103–107, https://doi.org/10.1126/science.282.5386.103.
[11] A.S. Perelson, Viral kinetics and mathematical models, The American Journal of Medicine 107 (6, Supplement 2) (1999) 49–52, https://doi.org/10.1016/S0002-9343(99)00382-4.
[12] H. Dahari, R.M. Ribeiro, C.M. Rice, A.S. Perelson, Mathematical modeling of subgenomic hepatitis C virus replication in Huh-7 cells, Journal of Virology 81 (2) (2007) 750–760, https://doi.org/10.1128/JVI.01304-06.
[13] M. Binder, N. Sulaimanov, D. Clausznitzer, M. Schulze, C.M. Hüber, S.M. Lenz, J.P. Schlöder, M. Trippler, R. Bartenschlager, V. Lohmann, L. Kaderali, Replication vesicles are load- and choke-points in the hepatitis C virus lifecycle, PLoS Pathogens 9 (8) (2013) 1–21, https://doi.org/10.1371/journal.ppat.1003561.
[14] J. Nakabayashi, A compartmentalization model of hepatitis C virus replication: an appropriate distribution of HCV RNA for the effective replication, Journal of Theoretical Biology 300 (2012) 110–117, https://doi.org/10.1016/j.jtbi.2012.01.023.
[15] A.K. McLean, F. Luciani, M.M. Tanaka, Trade-offs in resource allocation in the intracellular life-cycle of hepatitis C virus, Journal of Theoretical Biology 267 (4) (2010) 565–572, https://doi.org/10.1016/j.jtbi.2010.09.031.
[16] P. Padmanabhan, U. Garaigorta, N.M. Dixit, Emergent properties of the interferon-signalling network may underlie the success of hepatitis C treatment, Nature Communications 5 (May) (2014) 3872, https://doi.org/10.1038/ncomms4872.
[17] M. Castano-Arcila, A. Ramirez-Hurtado, C. Galvéz-de-León, J. Rodríguez-González, Dynamic analysis and control of the hepatitis C virus, in: 21st IFAC World Congress, 2020.
[18] A. Ay, D.N. Arnosti, Mathematical modeling of gene expression: a guide for the perplexed biologist, Critical Reviews in Biochemistry and Molecular Biology 46 (2) (2011) 137–151, https://doi.org/10.3109/10409238.2011.556597.
[19] S. Naylor, J.Y. Chen, Unraveling human complexity and disease with systems biology and personalized medicine, Personalized Medicine 7 (3) (2010) 275–289, https://doi.org/10.2217/pme.10.16.
[20] A. Cann, Replication of viruses, in: B.W. Mahy, M.H. Van Regenmortel (Eds.), Encyclopedia of Virology, 3rd edition, Academic Press, Oxford, 2008, pp. 406–412.
[21] G. Tavernier, O. Andries, J. Demeester, N.N. Sanders, S.C. De Smedt, J. Rejman, mrna as gene therapeutic: how to control protein expression, Journal of Controlled Release 150 (3) (2011) 238–247, https://doi.org/10.1016/j.jconrel.2010.10.020.
[22] M. Cobb, 60 years ago, Francis Crick changed the logic of biology, PLoS Biology 15 (9) (2017) 1–8, https://doi.org/10.1371/journal.pbio.2003243.
[23] B. Alberts, A. Johnson, J. Lewis, M. Raff, K. Roberts, P. Walter, Molecular Biology of the Cell, Garland Science, New York, 2002.
[24] A.L. Lehninger, D.L. Nelson, M.M. Cox, Lehninger Principles of Biochemistry, Worth Publishers, New York, 2000.
[25] Z. Szallasi, J. Stelling, V. Periwal, System Modeling in Cellular Biology: From Concepts to Nuts and Bolts, The MIT Press, Cambridge Massachusetts, London England, 2006.
[26] C. Cosentino, D. Bates, Feedback Control in Systems Biology, CRC Press Taylor and Francis Group, 2011.
[27] M. Moser, O. Leo, Key concepts in immunology, in: Vaccines Educational Supplement, Vaccine 28 (2010) C2–C13, https://doi.org/10.1016/j.vaccine.2010.07.022.
[28] J.S. Marshall, R. Warrington, W. Watson, H.L. Kim, An introduction to immunology and immunopathology, Allergy, Asthma and Clinical Immunology 14 (49) (2018) 1–14, https://doi.org/10.1186/s13223-018-0278-1.
[29] G. Prelich, Gene overexpression: uses, mechanisms, and interpretation, Genetics 190 (3) (2012) 841–854, https://doi.org/10.1534/genetics.111.136911.

[30] K. Bloom, F. van den Berg, P. Arbuthnot, Self-amplifying RNA vaccines for infectious diseases, Gene Therapy 28 (3) (2021) 117–129, https://doi.org/10.1038/s41434-020-00204-y.
[31] M.A. Morgan, M. Galla, M. Grez, B. Fehse, A. Schambach, Retroviral gene therapy in Germany with a view on previous experience and future perspectives, Gene Therapy 28 (1) (2021) 1–19, https://doi.org/10.1038/s41434-021-00237-x.
[32] N. Pardi, M.J. Hogan, F.W. Porter, D. Weissman, mRNA vaccines a new era in vaccinology, Nature Reviews Drug Discovery 17 (1) (2018) 261–279, https://doi.org/10.1038/nrd.2017.243.
[33] H.M. Sauro, Enzyme Kinetics for Systems Biology, Ambrosius Publishing, 2011.
[34] M. Ullah, O. Wolkenhauer, Stochastic Approaches for Systems Biology, Springer, 2011.
[35] M.B. Elowitz, A.J. Levine, E.D. Siggia, P.S. Swain, Stochastic gene expression in a single cell, Science 297 (5584) (2002) 1183–1186, https://doi.org/10.1126/science.1070919.
[36] N. Friedman, L. Cai, X.S. Xie, Linking stochastic dynamics to population distribution: an analytical framework of gene expression, Physical Review Letters 97 (16) (2006) 168302, https://doi.org/10.1103/PhysRevLett.97.168302.
[37] L.K. Nguyen, D. Kulasiri, Distinct noise-controlling roles of multiple negative feedback mechanisms in a prokaryotic operon system, IET Systems Biology 5 (2) (2011) 145–156, https://doi.org/10.1049/iet-syb.2010.0020.
[38] S.J. Park, S. Song, G.S. Yang, P.M. Kim, S. Yoon, J.H. Kim, J. Sung, The Chemical Fluctuation Theorem governing gene expression, Nature Communications 9 (1) (2018) 1–12, https://doi.org/10.1038/s41467-017-02737-0.
[39] D.T. Gillespie, A general method for numerically simulating the stochastic time evolution of coupled chemical reactions, Journal of Computational Physics 22 (4) (1976) 403–434, https://doi.org/10.1016/0021-9991(76)90041-3.
[40] D.T. Gillespie, Exact stochastic simulation of coupled chemical reactions, The Journal of Physical Chemistry 81 (25) (1977) 2340–2361, https://doi.org/10.1021/j100540a008.
[41] P.A. Iglesias, B.P. Ingalls, Control Theory and Systems Biology, The MIT Press, 2010.
[42] P.P. Dennis, H. Bremer, Modulation of chemical composition and other parameters of the cell at different exponential growth rates, EcoSal Plus 3 (1) (2008) 765–777, https://doi.org/10.1128/ecosal.5.2.3.
[43] M. Siwiak, P. Zielenkiewicz, Transimulation - protein biosynthesis web service, PLoS ONE 8 (9) (2013) e73943, https://doi.org/10.1371/journal.pone.0073943.
[44] R.M. Ribeiro, H. Li, S. Wang, M.B. Stoddard, G.H. Learn, B.T. Korber, T. Bhattacharya, J. Guedj, E.H. Parrish, B.H. Hahn, G.M. Shaw, A.S. Perelson, Quantifying the diversification of hepatitis C virus (HCV) during primary infection: estimates of the in vivo mutation rate, PLoS Pathogens 8 (8) (2012) e1002881, https://doi.org/10.1371/journal.ppat.1002881.
[45] L.V. Sharova, A.A. Sharov, T. Nedorezov, Y. Piao, N. Shaik, M.S. Ko, Database for mRNA half-life of 19 977 genes obtained by DNA microarray analysis of pluripotent and differentiating mouse embryonic stem cells, DNA Research 16 (1) (2009) 45–58, https://doi.org/10.1093/dnares/dsn030.
[46] J.E. Ladbury, S.T. Arold, Noise in cellular signaling pathways: causes and effects, Trends in Biochemical Sciences 37 (5) (2012) 173–178, https://doi.org/10.1016/j.tibs.2012.01.001.
[47] H. Youn, J.-K. Chung, Modified mRNA as an alternative to plasmid DNA (pDNA) for transcript replacement and vaccination therapy, Expert Opinion on Biological Therapy 15 (9) (2015) 1337–1348, https://doi.org/10.1517/14712598.2015.1057563, pMID: 26125492.
[48] C. Zhang, V. Tricou, G. Maruggi, H. Shan, J. Li, Advances in mRNA vaccines for infectious diseases, Frontiers in Immunology 10 (594) (2019) 1–13, https://doi.org/10.3389/fimmu.2019.00594.

[49] U. Garaigorta, F. Chisari, Hepatitis C virus blocks interferon effector function by inducing PKR phosphorylation, Cell Host Microbe 6 (6) (2009) 513–522, https://doi.org/10.1016/j.chom.2009.11.004.

[50] L.U. Aguilera, J. Rodriguez-Gonzalez, Studying HIV latency by modeling the interaction between HIV proteins and the innate immune response, Journal of Theoretical Biology 360 (2014) 67–77, https://doi.org/10.1016/j.jtbi.2014.06.025.

[51] L.U. Aguilera, J. Rodriguez-Gonzalez, Modeling the effect of tat inhibitors on HIV latency, Journal of Theoretical Biology 473 (2019) 20–27, https://doi.org/10.1016/j.jtbi.2019.04.018.

[52] C.-C. Shu, A. Chatterjee, G. Dunny, W.-S. Hu, D. Ramkrishna, Bistability versus bimodal distributions in gene regulatory processes from population balance, PLoS Computational Biology 7 (8) (2011) 1–13, https://doi.org/10.1371/journal.pcbi.1002140.

[53] B.S. Razooky, L.S. Weinberger, Mapping the architecture of the HIV-1 tat circuit: a decision-making circuit that lacks bistability and exploits stochastic noise, in: Methods Related to the Molecular Virology of HIV, Methods 53 (1) (2011) 68–77, https://doi.org/10.1016/j.ymeth.2010.12.006.

[54] A. Srikiatkhachorn, A. Mathew, A. Rothman, Immune-mediated cytokine storm and its role in severe Dengue, Seminars in Immunopathology 39 (1) (2017) 563–574, https://doi.org/10.1007/s00281-017-0625-1.

[55] A. Raychoudhuri, S. Shrivastava, R. Steele, H. Kim, R. Ray, R.B. Ray, ISG56 and IFITM1 proteins inhibit hepatitis C virus replication, Journal of Virology 85 (24) (2011) 12881–12889, https://doi.org/10.1128/JVI.05633-11.

[56] P.J. Hertzog, B.R. Williams, Fine tuning type I interferon responses, Cytokine and Growth Factor Reviews 24 (3) (2013) 217–225, https://doi.org/10.1016/j.cytogfr.2013.04.002.

CHAPTER

10

Impulsive nonlinear MPC with application to oncolytic virus therapy

María F. Villa-Tamayo[a], Anet J.N. Anelone[b], and Pablo S. Rivadeneira[a]

[a]Universidad Nacional de Colombia, Facultad de Minas, Grupo GITA, Medellín, Colombia, [b]School of Mathematics and Statistics, University of Sydney, Camperdown, NSW, Australia

10.1 Introduction

Oncolytic virus therapy (OVT) is a type of experimental treatment for cancer, which uses genetically modified viruses to specifically identify and attack tumor cells, without damaging healthy cells [4,6]. The idea behind OVT is that the virus infects a tumor cell and extensively replicates itself within it, eventually leading to the rupture of the affected cell (lysis) [14]. This process has several advantages such as self-propagation and infection of adjacent cells within the same tumor. In addition, viral infection during virotherapy leads to the activation of T cells against cancer neoantigens, i.e., it triggers immune responses against tumor cells [4,6].

Different works in the literature have also highlighted the safety related with the use of oncolytic viruses [3,4,6]. In this regard, OVT is considered as a treatment that could be used alone or in combination with surgery or other treatments to reduce tumors [3,4,6].

In the development of OVT there are different challenges to be addressed, such as determining the appropriate oncolytic virus for a specific cancer, strategies to increase the amount of virus that reaches the tumor as well as the virus effectiveness, and defining suitable protocols to administer viral doses [3,4,6]. To solve these situations, different points can be

Feedback Control for Personalized Medicine
https://doi.org/10.1016/B978-0-32-390171-0.00019-6

addressed, such as making superficial modifications of the virus or using superficial layers in it and implementing control formulations to establish suitable protocols. Regarding the first point, OVT has seen benefits from advances in virus engineering. Different types of oncolytic viruses have been genetically designed to be avirulent in healthy cells and to target specific tumor cells [4,6]. Some modifications have been made to improve virus survival against the immune system. For instance, the adenovirus has been modified with the nonimmunogenic polymer polyethylene glycol (PEG). Despite the fact that PEG manages to increase the retention time of the virus in the blood, the efficiency of the infection is reduced, since it makes it difficult for the virus to bind with cancer cells. Therefore, the virus has also been conjugated with the antibody herceptin [4], which is a monoclonal antibody capable of recognizing and binding to the Her2/neu protein regularly used in targeted therapy treatments. Thus, the so-called Ad-PEG-HER is generated, which is capable of specifically attacking cancer cells, with replication restricted only to Her2/neu-positive cells, capable of maintaining low levels of toxicity during their operation and with a longer retention time in the blood [4].

To better understand the effects of adenovirus in tumor cells and to propose different protocols for OVT, mathematical models have been developed [1,2,5]. Based on the experimental results presented in [4], in which three injections of 1×10^{10} viral particles of Ad-PEG and Ad-PEG-HER were applied on days 0, 2, and 4 to nude mice with tumor of sizes between 100 mm^3 and 120 mm^3, the work in [1] introduced a mathematical model for OVT. This model can be used as an in silico platform to represent the evolution of the tumor volume with the corresponding biological parameters that adjust the model to the experimental data in [4], which in turn allows investigating possible therapy protocols [1].

To determine adequate viral doses and improve the efficacy of the therapy, approaches from control theory have been introduced [16,17]. As detailed in [16], OVT can be analyzed under the scope of impulsive control systems (ICSs). These type of systems are characterized by a forced response when a control input of short duration with respect to the sampled time is applied and a free response during the remaining sample time when the input is null [7]. Thus, OVT fits under this description as the considered input is the virus load applied as an injection on predefined days and the remaining time no input is delivered. Two types of control strategies have been explicitly used for OVT as an ICS: a feedback control in [16] to explore different protocols and the advantages of control theory and a model predictive control (MPC) in [17] to find the optimal sequence of system inputs (the virus load) that minimizes the distance of the number of tumor cells to a target and the magnitude of the input. In addition, in the literature other impulsive control strategies have been applied to different biomedical processes. For instance, strategies for type 1 diabetes and influenza treatment can be found in [8,9,14] and [10], respectively, and protocols for HIV treatment have been proposed in [11,12].

In this chapter, a personalized protocol is presented by using the experimental and mathematical modeling work in [1,4,5] and the formulation of an impulsive MPC to obtain the viral schedule to reduce tumor cells. Due to nonlinearities in the model that represent the dynamics of tumor cells and the virus, the prediction model in the MPC is updated at each time step, so that an affine linear approximation of the system is used around the current operating point. The results presented consider the oncolytic virus Ad-PEG-HER and the parameters of the five nude mice identified from [4]. Furthermore, the injection of viral particles is modeled as an impulse applied each 2 days, as in [17]. The results of the proposed nonlinear MPC are compared with the experimental protocol used in [4]. The MPC performance is also evaluated in the presence of simultaneous variations in the biological rates.

The outline of the chapter is as follows. In Section 10.2, the notation used in the chapter is introduced. In Section 10.3, the model of the interactions between the oncolytic virus and tumor cells is described with its representation as an ICS. In Section 10.4, the MPC formulation is detailed. In Section 10.5, the results are presented. Finally, the conclusions and perspectives are exposed in Section 10.6.

10.2 Notation

The sets of reals, n-dimensional column vectors, and matrices of dimension $n \times m$ are denoted as, $\mathbb{R}$, $\mathbb{R}^n$, and $\mathbb{R}^{n\times m}$, respectively; I_n denotes the identity matrix of dimension $n \times n$. The transpose of a matrix A is represented as A'. Given $x \in \mathbb{R}^n$ and a matrix $Q \geq 0 \in \mathbb{R}^{n\times n}$, $\|x\|_Q^2 = x'Qx$.

10.3 Model of oncolytic virus therapy

The model used in here as an in silico platform to test the control formulation is the one introduced in [1] and improved in [5]. This model consists of three ordinary differential equations to represent the dynamics of a tumor-viral treatment as follows:

$$\begin{aligned}
\frac{dS(t)}{dt} &= r log\left(\frac{L}{S(t)}\right) S(t) - \frac{\beta S(T)V(t)}{S(t)+I(t)+\epsilon}, \\
\frac{dI(t)}{dt} &= \frac{\beta S(t)V(t)}{S(t)+I(t)+\epsilon} - d_I I(t), \\
\frac{dV(t)}{dt} &= u(t) - d_V V(t) + \alpha d_I I(t),
\end{aligned} \tag{10.1}$$

where the three state variables S, I, and V correspond to the density of susceptible tumor cells, the density of virus-infected tumor cells, and the density of virus particles at the tumor site relative to the number of cells, respectively; $T = S + I$ corresponds to the total tumor cell population and the output of the model. The input of the model is $u(t)$, which represents the virus injection with amplitude u_v on predefined days τ_k. Thus, it has the form

$$u(t) = \begin{cases} u_v, & t = \tau_k,\ k \in \mathbb{N}, \\ 0, & \text{otherwise.} \end{cases} \tag{10.2}$$

In the model we assume a single average infectivity rate β [day^{-1}] and a single viral decay rate d_V [day^{-1}], i.e., it does not consider modifications or conjugations in the virus via replication within the tumor cells. The tumor growth is described by a Gompertz function $r log\left(\frac{L}{S(t)}\right) S(t)$, with L [cells$\times 10^6$] denoting the carrying capacity of the tumor and r [day^{-1}] being the tumor growth rate. The death rate of infected tumor cells is represented by d_I [day^{-1}] and the viral burst size by α [virus$\times 10^9$]. Parameter ϵ is small and has been included to avoid the singularity when $S + I \to 0$. Since nude mice have a genetic mutation which suppresses the immune system, immune response mechanisms were not included in the model [1,4].

The model fit well the data from [4] showing the time course of total number of tumor cells in nude mice during OVT [1]. The parameters and initial conditions for each subject are presented in Table 10.1.

TABLE 10.1 Initial conditions and parameter values of the model (10.1).

Symbol	Unit	S1	S2	S3	S4	S5
S_0	cells	238.3535	200.0340	101.5400	140.3436	128.1481
I_0	cells	0	0	0	0	0
V_0	virus	0	0	0	0	0
r	day^{-1}	0.0358	0.0702	0.0225	0.0314	0.0586
L	cells$\times 10^6$	6518.7	3200.9	4910.6	8839.8	996.6239
β	day^{-1}	0.7055	0.9300	0.2000	1.3039	1.4399
d_I	day^{-1}	2	1.9922	2	0.1	0.1000
d_v	day^{-1}	2.3042	3.4907	3.4999	1.8692	1.8196
α	virus$\times 10^9$	3.5	3.5	3.5	3.5	3.5

10.3.1 Impulsive representation of the model

As the virus is administered to the subject by injection every few days, the model for OVT (10.1) can be represented as an ICS. This because the input to the system (the viral injection) is of very short duration in relation to the sampling time (here an application of the virus every 2 days is considered). Therefore, the system can be represented by a forced response in some impulsive moments when the virus is applied to the subject and a free response during the remaining time when the input is null.

To represent the OVT model as an ICS, let us define the state $x(t) = [S(t) \;\; I(t) \;\; V(t)]'$. Then, the nonlinear system (10.1) is rewritten as

$$\begin{aligned} \dot{x}(t) &= f(x(t)), && x(0) = x_0, \; t \neq \tau_k, \\ x(\tau_k^+) &= x(\tau_k) + Bu(\tau_k), && t = \tau_k, \quad k \in \mathbb{N}, \\ T(\tau_k) &= Cx(\tau_k), && t = \tau_k, \quad k \in \mathbb{N}, \end{aligned} \tag{10.3}$$

where $f(x)$ denotes the autonomous response of the system, $B = [0 \;\; 0 \;\; 1]'$ is the constant matrix that represents the effect of viral injections at times τ_k, $k = 1, 2, ...$, over the state x, and $C = [1 \;\; 1 \;\; 0]$ is the constant matrix of the output equation. An approximation of the nonlinear system can be obtained by linearizing the free response $f(x)$ around an operation point (x_{op}, u_{op}). This point is not necessarily an equilibrium of the system. Then, as $\dot{x}(t) \approx f(x_{op}) + A(x(t) - x_{op})$ the system is described as

$$\begin{aligned} \dot{x}(t) &= Ax(t) + E, && x(0) = x_0, \; t \neq \tau_k, \\ x(\tau_k^+) &= x(\tau_k) + Bu(\tau_k), && t = \tau_k, \quad k \in \mathbb{N}, \\ T(\tau_k) &= Cx(\tau_k), && t = \tau_k, \quad k \in \mathbb{N}, \end{aligned} \tag{10.4}$$

where matrix A is the Jacobian of the system with respect to the operation point and matrix E is the term defined as $E = f(x_{op}) - Ax_{op}$.

In addition, system (10.4) can be described in times τ_k by means of an underlying discrete system of the form

$$x(k+1) = A_d x(k) + B_d u(k) + E_d, \tag{10.5}$$

with $A_d = e^{AT_s}$, $B_d = e^{AT_s}B$, $E_d = \int_0^{T_s} e^{As} ds E$, $x(0) = x(\tau_0)$, and T_s a fixed sampling time defined as $T_s = \tau_{k+1} - \tau_k$. This discrete representation is the one used in the MPC formulation in the next section. A complete characterization of ICS and the underlying discrete system can be seen in [13].

10.4 Model predictive control for oncolytic virus therapy

The control objective is to reduce the total number of tumor cells T below 50 in less than 60 days by conducting OVT based on control theory. To achieve this, the block diagram shown in Fig. 10.1 is proposed as an initial approach, where the viral therapy is posed as a closed-loop system by using the tumor information at each time step to calculate, by means of a controller, the appropriate viral dose to meet the objectives.

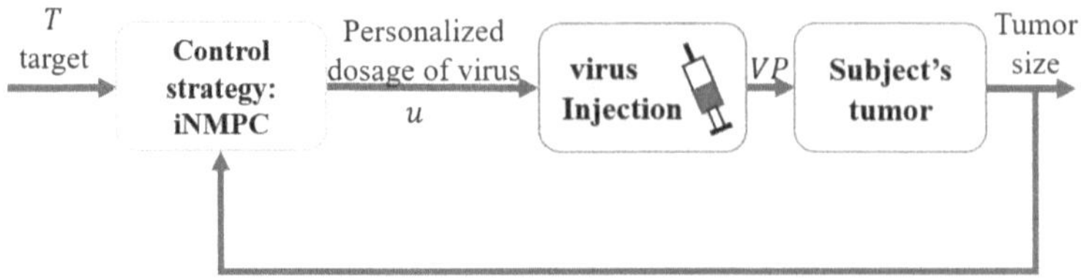

FIGURE 10.1 Impulsive model predictive control (iNMPC) for oncolytic virus therapy.

Since one of the biggest concerns is the toxicity caused by the accumulation of virus in the system [4], the goal is to achieve tumor regression with the minimum viral doses. To that end, the impulsive nonlinear MPC (iNMPC) strategy is proposed to find the sequence of injections to be applied such that both the difference between the system output and the target and the amount of virus administered are minimized. Thus, the optimization problem to solve at each step time k is

$$\begin{aligned} \min_{\mathbf{U}} V_N &= \sum_{j=0}^{H_p-1} \|T(j) - T_{ref}\|_Q^2 + \sum_{j=0}^{H_u-1} \|u(j)\|_R^2 \\ s.t. & \\ & x(j+1) = f_d(x(j), u(j)), \\ & T(j) = Cx(j), \\ & x(j) \in \mathcal{X}, \quad u(j) \in \mathcal{U}, \end{aligned} \tag{10.6}$$

where $\mathcal{X}$ is the set of state constraints $\mathcal{X} = \{x | x \geq [0, 0, 0]'\}$ and $\mathcal{U}$ is the set of input constraints $\mathcal{U} = \{u | u \geq 0\}$. The solution of the problem is the optimal sequence $U^* = \{u^*(k), ..., u^*(k + H_u - 1), ..., u^*(k + H_p - 1)\}$, with $u^*(k + H_u - 1) = \cdots = u^*(k + H_p - 1)$, from which the first element $u^*(k)$ is applied to the plant.

10.4.1 Matrix representation of the cost function

The optimization problem can be rewritten by replacing the nonlinear function with the discrete linear system (10.5). With the purpose of having a better approximation of the system at each time k, the matrix A is updated at each step time with the current operation state x_{op} [15].

Next, the following expansion of the state in the prediction horizon is considered:

$$X = A_{au}x(k) + B_{au}U + E_{au}, \quad Y = C_{au}X, \tag{10.7}$$

$$
\begin{bmatrix} x(k+1) \\ \vdots \\ x(k+Hu) \\ \vdots \\ x(k+Hp) \end{bmatrix} = \begin{bmatrix} A_d \\ \vdots \\ A_d^{H_u} \\ \vdots \\ A_d^{H_p} \end{bmatrix} x(k)
+ \begin{bmatrix} B_d & \cdots & 0 \\ \vdots & \ddots & \vdots \\ A_d^{H_u-1} B_d & \cdots & B_d \\ \vdots & \ddots & \vdots \\ A_d^{H_p-1} B_d & \cdots & \sum_{i=0}^{H_p-H_u} A_d^i B_d \end{bmatrix} \begin{bmatrix} u(k) \\ \vdots \\ u(k+H_u-1) \end{bmatrix}
+ \begin{bmatrix} E_d \\ \vdots \\ \sum_{i=0}^{H_p-1} A_d^i E_d \end{bmatrix},
$$

$$
\begin{bmatrix} T(k+1) \\ \vdots \\ T(k+Hp) \end{bmatrix} = \begin{bmatrix} C & \cdots & 0 \\ \vdots & \ddots & \vdots \\ 0 & \cdots & C \end{bmatrix} X. \tag{10.8}
$$

When rewriting the cost function as $V = V1 + V2$ with

$$
\begin{aligned}
V1 &= (C_{au}X - I_{H_p,n_y}T_{ref})'\tilde{Q}(C_{au}X - I_{H_p,n_y}T_{ref}) \\
&= (C_{au}(A_{au}x(k) + E_{au}) - I_{H_p,n_y}T_{ref} + C_{au}B_{au}U)'\tilde{Q}(\cdot) \\
&= (a + bU)'\tilde{Q}(a + bU), \\
V2 &= (U)'\tilde{R}(U) = (cU)'\tilde{Q}(cU),
\end{aligned} \tag{10.9}
$$

the quadratic expression of the cost is obtained:

$$
V = \frac{1}{2}U'HU + f'U + cte,
$$

with $H = 2(b'\tilde{Q}b + c'\tilde{R}c)$, $f = 2(a'\tilde{Q}b)'$, $cte = a'\tilde{Q}a$, $a = C_{au}(A_{au}x(k) + E_{au}) - IT_{ref}$, $b = C_{au}B_{au}$, and $c = [I_{H_u,n_u}]$. The matrix $I_{z,v}$ is the identity matrix I_v expanded z times, and matrices $\tilde{Q}$ and $\tilde{R}$ are

$$
\tilde{Q} = \begin{bmatrix} CQ & \cdots & 0 \\ \vdots & \ddots & \vdots \\ 0 & \cdots & CQ \end{bmatrix}, \quad \tilde{R} = \begin{bmatrix} R & \cdots & 0 \\ \vdots & \ddots & \vdots \\ 0 & \cdots & R \end{bmatrix}.
$$

10.4.2 Matrix representation of the constraints

System constraints include both control input constraints and state constraints of the form

$$u_{min} \leq u(j) \leq u_{max}, \quad x_{min} \leq x(j) \leq x_{max}, \tag{10.10}$$

which must be satisfied throughout the control and prediction horizons, respectively. Therefore, the respective expansions of these constraints are

$$\begin{aligned} LB_u \leq U \leq UB_u, \\ A_{au}x(k) + B_{au}U + E_{au} \leq X_{max}, \\ Xmin \leq A_{au}x(k) + B_{au}U + E_{au}, \end{aligned} \tag{10.11}$$

with

$$LBu = \begin{bmatrix} u_{min} \\ \vdots \\ u_{min} \end{bmatrix}, \quad UBu = \begin{bmatrix} u_{max} \\ \vdots \\ u_{max} \end{bmatrix}, X_{max} = \begin{bmatrix} x_{max} \\ \vdots \\ x_{max} \end{bmatrix}, \quad X_{min} = \begin{bmatrix} x_{min} \\ \vdots \\ x_{min} \end{bmatrix}.$$

State constraints can be rewritten in the form $A_{in}u \leq b_{in}$, resulting in

$$A_{in} = \begin{bmatrix} B_{au} \\ -B_{au} \end{bmatrix}, \quad b_{in} = \begin{bmatrix} -A_{au}x(k) - E_{au} + X_{max} \\ A_{au}x(k) + E_{au} - X_{min} \end{bmatrix}. \tag{10.12}$$

Given sets $\mathcal{X}$, $\mathcal{U}$, the lower bounds for the state and the input are established as $u_{min} = 0$, $x_{min} = [0 \ \ 0 \ \ 0]'$, and the upper bounds can be defined using therapeutic constraints or simply left as empty sets.

It should be emphasized that the increased matrices A_{au}, E_{au} must be recalculated at each time step k due to the update of matrices A and E at the operation point x_{op}. In turn, the inequality constraint matrices on the state must be calculated at each time step.

A summary of the steps required to set the MPC strategy is presented in Algorithm 10.1. These steps and the expansion of the matrices are suitable to reproduce the results obtained in the next section. The reader is encouraged to write to the authors to ask any questions about the code.

10.5 Results

This section presents and discusses computer simulations showing the performance of the proposed nonlinear MPC strategy for OVT. The simulations represent the treatment of solid tumors in nude mice with the oncolytic Ad-PEG-HER virus, as in [4]. First, the antitumor efficacy of the iNMPC is compared to the experimental protocol from [4] which consists

Algorithm 10.1: Algorithm for the nonlinear MPC to conduct oncolytic virus therapy

In summary, the nonlinear MPC for OVT is formulated as:

1. Define control parameters Q, R, H_p, H_u.
2. Set system constraints u_{min}, u_{max}, x_{min}, x_{max}.
3. Expand the constraints on the decision variable LB_u and UB_u.
4. At each time step k, calculate matrices A, E to obtain the approximation of the nonlinear impulsive system around the current operation point x_{op}.
5. Calculate the underlying discrete system to obtain matrices A_d, B_d, C_d, and E_d.
6. Expand the matrices on the prediction horizon to get A_{au}, B_{au}, E_{au}, and C_{au}.
7. Compute H and f of the cost function and A_{in}, b_{in} for state constraints.
8. Solve the constrained quadratic problem to find the sequence of inputs U^*.
9. Apply the first element of the sequence $u^*(k)$ to the plant (the tumor in the subject) and repeat from step 4.

of three injections of 10^{10} virus particles on days 0, 2, and 4. Recall that the control objective is to steer and keep the total number of tumor cells in the therapeutic zone $T < 50$. The nonlinear MPC is configured with a sampling time $T_s = 2$ days and a reference $T_{ref} = 0$. As observed in Fig. 10.2, the iNMPC has a higher antitumor efficacy than the experimental protocol. Unlike the experimental protocol, the iNMPC achieves the control objectives. Although the total dose applied with the nonlinear MPC is higher than the total dose of the experimental protocol, the nonlinear MPC delivers the minimum viral doses required to steer the tumor to the therapeutic zone.

Next, additional simulations are conducted to investigate the robust performance of the nonlinear MPC. The robustness of the iNMPC is evaluated to acknowledge parameter uncertainty between the plant and the model. To that end, the parameters of the plant are varied by 10% of their nominal values. The performance of the nonlinear MPC is shown in Figs. 10.3 and 10.4. The iNMPC manages to steer the tumor to the therapeutic zone despite the plant–model mismatch (see Fig. 10.3). It is observed that, to compensate for the mismatch, the MPC computes higher total viral doses than the nominal case. In addition, the growth of total tumor cells is due to infected cells because susceptible tumor cells are reduced in a very short time (see Fig. 10.4). Tumor regression results from the loss of infected cells due to the lysis process caused by the virus.

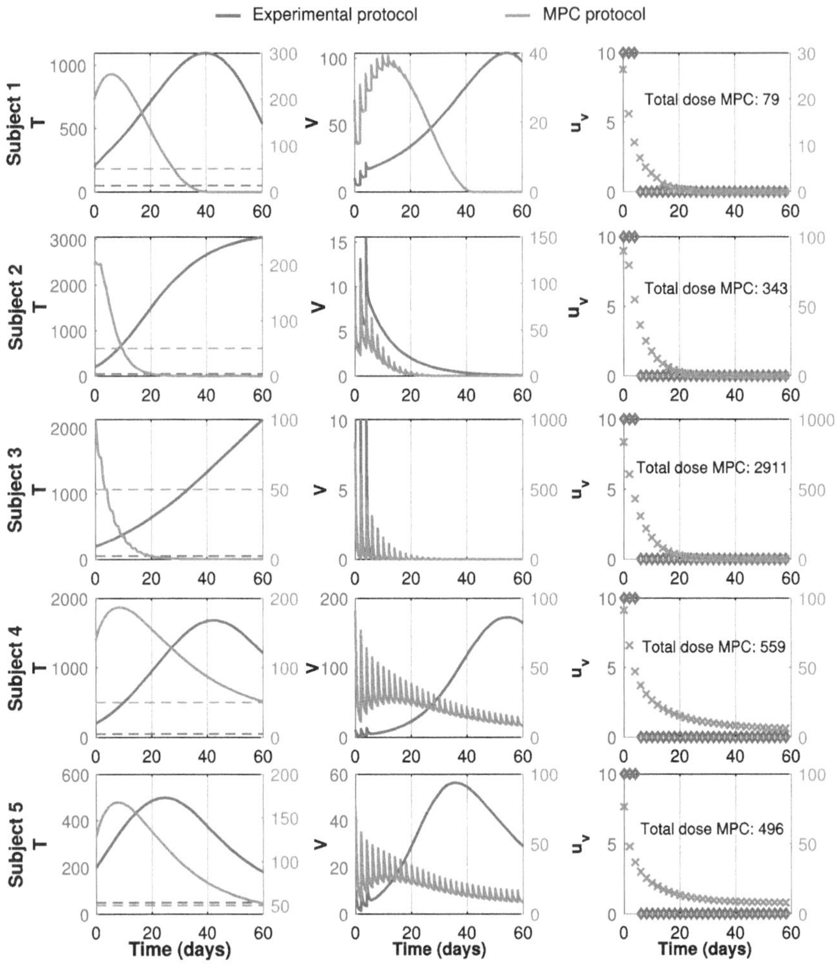

FIGURE 10.2 Comparison between the system evolution under the experimental protocol (blue (dark gray in print version) lines and left axes) and under the MPC strategy (red (gray in print version) lines and right axes). Dashed lines in the first column refer to the target zone. Total dose applied with the MPC is depicted in the right column. For all subjects the total dose with the experimental protocol is 30.

Together, these results suggest that the proposed iNMPC provides an analytical framework to deliver personalized, optimal, and robust control strategies which may be beneficial to OVT.

10.6 Conclusion

This chapter proposed an iNMPC with application to OVT. This scheme has been formulated using an underlying discrete-time linear subsystem

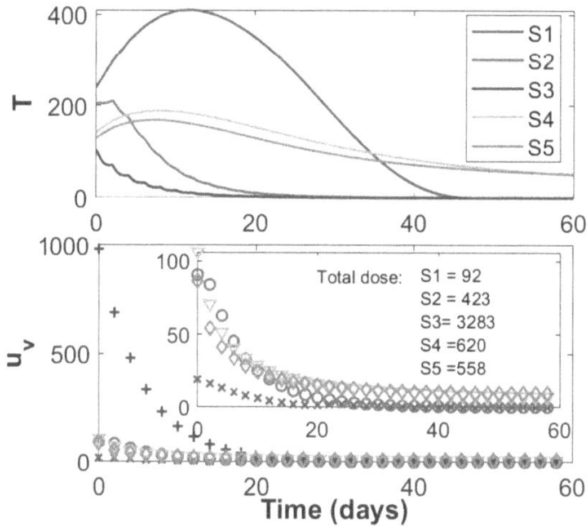

FIGURE 10.3 System evolution when applying the iNMPC controller and there are simultaneous variations of 10% in parameters r, d_i, d_v, and α of the plant (the subject). The upper row shows the time course of the total number of tumor cells for each subject, the green (mid gray in print version) zone is the target $T < 50$ cells. The bottom row shows the time course of viral injections and the total dose administered for each subject.

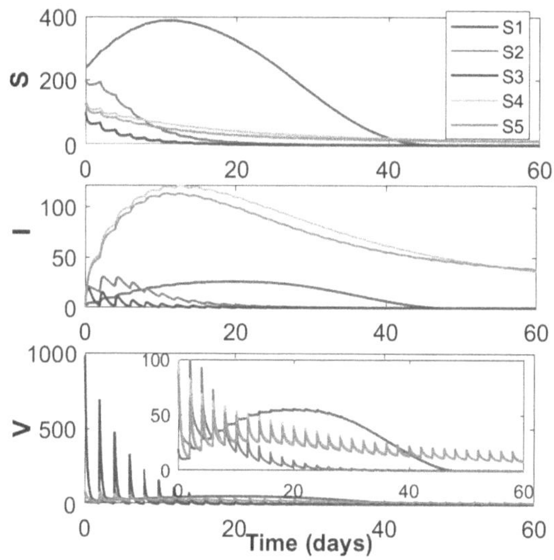

FIGURE 10.4 State evolution of each subject when applying the MPC control strategy and considering simultaneous variations of 10% in parameters r, d_i, d_v, and α of the plant. The upper row shows susceptible tumor cells S, the middle row shows infected cells I, and the bottom row shows virus particles in vivo V.

with matrices that approximate the nonlinear behavior at each time step. The nonlinear MPC has been shown to calculate personalized viral injections to achieve tumor regression in a therapeutic target zone. When considering a plant–model mismatch caused by parameter variations, the MPC increases the viral dose to successfully reduce total tumor cells.

For further works, a state observer should be included to estimate the complete state from the output T, to implement the control strategy. In addition, the treatment could be improved by adding an extra system equation or a constraint related to virus toxicity, so that the controller is formulated to minimize it.

References

[1] A.L. Jenner, C.O. Yun, P.S. Kim, A.C.F. Coster, Mathematical modelling of the interaction between cancer cells and an oncolytic virus: insights into the effects of treatment protocols, Bull. Math. Biol. 80 (6) (2018) 1615–1629.
[2] N.L. Komarova, D. Wodarz, ODE models for oncolytic virus dynamics, J. Theor. Biol. 263 (4) (2010) 530–543.
[3] A. Haseley, C. Alvarez-Breckenridge, A.R. Chaudhury, B. Kaur, Advances in oncolytic virus therapy for glioma, Recent Pat. CNS Drug Discov. 4 (2009) 1–13.
[4] P. Kim, J. Sohn, J. Choi, et al., Active targeting and safety profile of PEG-modified adenovirus conjugated with herceptin, Biomaterials 32 (9) (2011) 2314–2326.
[5] A.L. Jenner, P.S. Kim, F. Frascoli, Oncolytic virotherapy for tumours following a Gompertz growth law, J. Theor. Biol. 480 (2019) 129–140.
[6] K. Twumasi-Boateng, J.L. Pettigrew, Y.Y.E. Kwok, J.C. Bell, B.H. Nelson, Oncolytic viruses as engineering platforms for combination immunotherapy, Nat. Rev. Cancer 18 (7) (2018) 419–432.
[7] P.S. Rivadeneira, C.H. Moog, Impulsive control of single-input nonlinear systems with application to HIV dynamics, Appl. Math. Comput. 218 (2012) 8462–8474.
[8] P.S. Rivadeneira, A. Ferramosca, A.H. González, Impulsive zone model predictive control with application to type I diabetic patients, in: 2016 IEEE Conf. on Control Appl. (CCA), 2016, pp. 544–549.
[9] M.F. Villa Tamayo, M.A. Caicedo Álvarez, P.S. Rivadeneira, Handling parameter variations during the treatment of type 1 diabetes mellitus: in silico results, Math. Probl. Eng. (2019), https://doi.org/10.1155/2019/2640405.
[10] A. Boianelli, N. Sharma-Chawla, D. Bruder, E.A. Hernandez-Vargas, Oseltamivir PK/PD modelling and simulation to evaluate treatment strategies against Influenza-Pneumococcus coinfection, Front. Cell Infect. Microbiol. 6 (60) (2016) 1–11.
[11] A.J.N. Anelone, S.K. Spurgeon, Prediction of the containment of HIV infection by antiretroviral therapy – a variable structure control approach, IET Syst. Biol. 11 (1) (2016) 44–53, https://doi.org/10.1049/iet-syb.2016.0028.
[12] A.J.N. Anelone, S.K. Spurgeon, Modelling and simulation of the dynamics of the antigen-specific T cell response using variable structure control theory, PLoS ONE 11 (11) (2016) 1–22, https://doi.org/10.1371/journal.pone.0166163.
[13] P.S. Rivadeneira, A. Ferramosca, A.H. González, Control strategies for non-zero setpoint regulation of linear impulsive systems, IEEE Trans. Autom. Control 63 (9) (2018) 2994–3001.
[14] M.F. Villa-Tamayo, M.A. Caicedo, P.S. Rivadeneira, Offset-free MPC strategy for nonzero regulation of linear impulsive systems, ISA Trans. (2020), https://doi.org/10.1016/j.isatra.2020.01.005.
[15] J. Espinosa, J. Vandewalle, V. Wertz, Fuzzy logic, identification and predictive control, in: Advances in Industrial Control, London, 1st edn., 2010.

[16] A.J.N. Anelone, M.F. Villa-Tamayo, P.S. Rivadeneira, Oncolytic virus therapy benefits from control theory, R. Soc. Open Sci. 7 (7) (2020) 200473, https://doi.org/10.1098/rsos.200473.

[17] M.F. Villa-Tamayo, A.J.N. Anelone, P.S. Rivadeneira, Tumor reduction using oncolytic viruses under an impulsive nonlinear estimation and predictive control scheme, IEEE Control Syst. Lett. 5 (5) (2021) 1705–1710, https://doi.org/10.1109/LCSYS.2020.3043185.

CHAPTER

11

Is the isolated heart a relaxation-oscillator?

Carlos Enrique Gálvez-de León, Amelia Ríos, Karen Garza Cuellar, Bruno A. Escalante, and Jesús Rodríguez-González

Center for Research and Advanced Studies of the National Polytechnic Institute, Monterrey Unit, Parque PIIT, Apodaca, NL, Mexico

11.1 Introduction

Physiological rhythms permeate biology on all levels, from molecular processes to complex population dynamics [1]. It is well known that the heart is a specialized muscle that contracts regularly and continuously, pumping blood to the body and the lungs. The heartbeat depends on electrical depolarization to trigger mechanical contraction. Each heartbeat results from electrical activity originating in a specialized pacemaker region called the sinoatrial node in the healthy heart. This electrical activity then spreads through the atria, then through the atrioventricular node, and the His–Purkinje system to produce orderly contractions of the atria and ventricles [2].

From the oscillation theory viewpoint, the heart can be treated as a self-sustained oscillator, where periodic electrical stimulation can be regarded as an external forcing [3]. Excitation–contraction coupling is an essential concept for understanding the synchronization phenomena at the whole heart scale. The phase resetting curve (PRC) is a handy tool for characterizing how oscillating systems respond to stimuli [4]. We use a model that portrays the functioning of the heart using the Van der Pol oscillator, and it was initially proposed by the Dutch electrical engineer and physicist Balthazar van der Pol while he was working at Philips. Van der Pol found stable oscillations, which he called relaxation-oscillations and are

Feedback Control for Personalized Medicine
https://doi.org/10.1016/B978-0-32-390171-0.00020-2

now known as limit cycles, in electrical circuits employing vacuum tubes. The Van der Pol equation is now a basic model for oscillatory processes in physics, electronics, biology, neurology, sociology, and economics [5].

Numerous theoretical and experimental investigations to explore whether biological oscillators correspond to relaxation-oscillations have been reported. Among many examples, phase resetting of a single neuron is theoretically analyzed by Best [6] and experimentally verified by Guttman and colleagues [7]. In addition, Somers and Kopell have proven that relaxation-oscillators have properties like they need just a few or even one cycle to synchronize, and the synchronization is stable in the presence of nonuniformity of natural frequencies [8–10]. However, the question whether the isolated heart behaves like a relaxation-oscillator has received little attention. This work is devoted to partially filling this gap in knowledge. In a rhythmically responding heart, a natural question is how electrical stimulus shifts the timing of the rhythm. In particular, we explored whether the heart behaves like a relaxation-oscillator when disturbed by an electrical stimulus of short duration at different times of the cardiac cycle. A graphical representation of the experimental setting is depicted in Fig. 11.1A, showing controlled pulse stimulation (input) and mechanical response variables (output).

11.2 Biological background

Excitation–contraction coupling is the process which translates electrical signals that originate from pacemaker cells and are distributed through the heart wall by Purkinje fibers to cardiomyocyte contraction, followed by subsequent relaxation [11].

Cardiac pressure–volume (P-V) loop analysis provides detailed information on cardiac function and is the golden standard for functional assessment and assessing a variety of experimentally induced pathologies. P-V loops represent the changes in pressure and volume of the left ventricle (LV) during one cardiac cycle [12,13]. Cardiac P-V analysis presents particular advantages over other measures of cardiac function, as it allows for measurement of ventricular function independent of loading conditions and heart rate. Fig. 11.1B illustrates a typical P-V loop. Systole occurs when the heart contracts to pump blood out, and diastole occurs when the heart relaxes after contraction. During diastole, the atrioventricular valves are open, and the aortic and pulmonary valves are closed. As a result, blood flows into the heart, filling the atria and ventricles, and pressure increases from d to a. Pressure then rises from a to b during iso-volumetric contraction and from b to c during ventricular ejection. Finally, at c, the aortic valves close, and pressure falls during isovolumetric relaxation from c back to d.

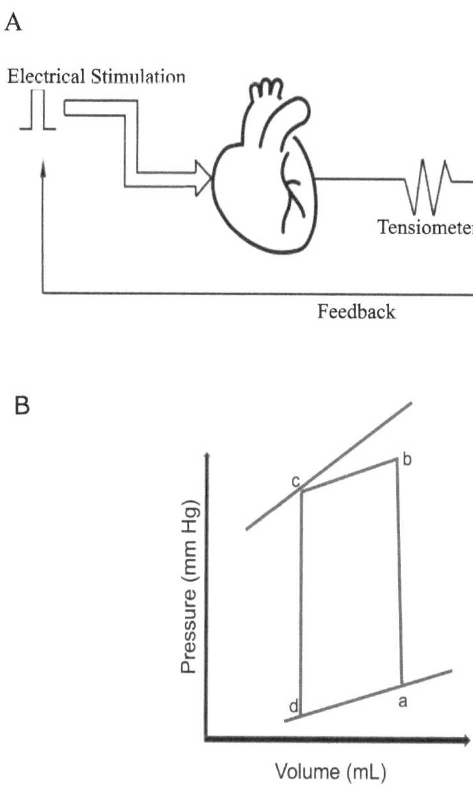

FIGURE 11.1 Graphical representation of the experimental setup. (A) Feedback activates a stimulus that is applied in a controlled phase. Contraction is registered in the computer. (B) Typical pressure–volume loop present in the left ventricle.

11.3 Theoretical background

The relaxation-oscillator has a venerable history. It was studied in the 19th century by Lord Rayleigh, analyzed further by Van der Pol, used by him and Van der Mark to model the heart, and claimed by the physiologist A.V. Hill to be the oscillator with which alone we are concerned in physiology [14]. Balthazar Van der Pol (1889–1959) was a Dutch electrical engineer who initiated modern experimental dynamics in the laboratory during the 1920s and 1930s. His studies of oscillation in an electronic circuit led to the derivation of the Van der Pol equation, a paradigmatic model of oscillation theory and nonlinear dynamics. The Van der Pol equation first appeared in his article on relaxation-oscillation published in the *Philosophical Magazine* in 1926 [15,16]. This equation is one of the most intensely studied systems in nonlinear dynamics, and it is widely associated with the very first development of a nonlinear oscillations theory [17]. The Van der Pol equation is now a basic model for oscillatory dynamics in physics, biology, neurology, and economics [18]. Van der Pol himself built many electronic circuit models of the human heart to study the range of stability of heart dynamics [19,20]. His investigations adding an external driving signal were analogous to how a pacemaker drives a real heart. Furthermore, the Van der Pol equation has been used to model coupled neurons [21]. The FitzHugh–Nagumo equation is a planar vector field that extends the Van der Pol equation as a model for the action potentials of neurons [22]. Every real physical system is nonlinear, displaying at least saturation behavior, and biological systems are particularly subject to nonlinearities.

According to nonlinear dynamics theory, a two-dimensional phase plane corresponds to a visual display between two state variables in a dynamic system. A stable limit cycle is a closed trajectory in phase space having the property that all the neighboring trajectories approach the limit cycle as the time approaches infinity. Fig. 11.1B shows a limit cycle used to model real-world oscillatory systems. Self-sustained rhythms are widespread in biology; examples include the heartbeat, respiration, hormonal cycles, circadian clocks, and calcium oscillations. Such self-sustained oscillations have stable limit cycles [23].

System identification is a well-established tool in engineering systems and control theory that allows constructing a predictive dynamic model of a biological system without any mechanistic details [24]. For linear systems, valuable information on the dynamical system is extracted by studying the response to a single pulse. A fundamental aspect of a dynamic process is the timescale on which it acts. The frequency response relies on frequency domain analysis, which applies only to linear systems. A system is linear if it acts additively: the response to two simultaneous perturbations is equivalent to the sum of the responses to the individual perturbations. Linearity is a helpful conceit; however, natural systems are perfectly nonlinear. In particular, biological systems typically involve saturation, rather than additivity of responses [25].

A phase response curve of an oscillating neuron measures the phase shifts in response to stimuli delivered at different times in its cycle. A PRC is obtained by systematically applying the same perturbation at different phases and measuring the resulting phase shifts. Note that PRC has some analogy to the impulse–response concept applied in linear systems [26]. The overwhelming complexity of analyzing the interaction between oscillators is reduced to few simple mathematical models for nonlinear systems. Instead of studying a mechanistic model, a single pulse-like perturbation of limit-cycle oscillators is used to analyze if entrainment can be induced [27].

A striking feature of PRCs is their universality. Despite enormous differences in timescales and oscillation-generating mechanisms, many properties appear quite similar [26]. Two different types of PRCs are obtained if we vary the perturbation strength. Type 1 PRCs show smooth curves, whereas discontinuous PRCs represent Type 0 PRC. Both types have been predicted in cardiology and circadian rhythms [28]. The cardiac conducting system can be treated as a network of self-excitatory elements: the sinoatrial node, the primary pacemaker, the atrioventricular node, and the His–Purkinje system. Since these elements exhibit oscillatory behavior, they can be modeled as nonlinear oscillators. We decided to use the Van der Pol model, a relaxation-oscillator, for three reasons. Firstly, this model adapts its intrinsic frequency to the frequency of the external driving signal without changing its amplitude. Secondly, a relation of the Van der Pol oscillator to the heartbeat was proposed by FitzHugh. Finally, the Van der Pol equation is a simplification of the Hodgkin and Huxley equations, describing the generation of action potentials [29,30].

11.4 Methodology

Male CD1 mice (weight, 30–40 g) were obtained from the Experimental Animal Care Center from Centro de Investigación y Estudios Avanzados del Instituto Politécnico Nacional (CINVESTAV-IPN), Mexico. All the procedures conformed to the National Institutes of Health Guide for the Care and Use of Laboratory Animals (1996) and were approved by the Institutional Ethics Review Committee for Animal Experimentation of CINVESTAV-IPN. The animals were acclimated for 1 week in metabolic cages (Nalgene, USA) and they were maintained in a temperature-controlled room (22°C) under a 12/12-h light/dark cycle.

11.4.1 Experimental procedure

A typical experiment was conducted as follows. The isolated heart was placed in a Langendorff system and electrically stimulated with two electrodes superficially attached close to the sinoatrial node. The electrodes were connected to the output of a stimulator. The heart rate (HR) and the intensity of the contractile force (ICF) were recorded as an approximation of the contraction force [31]. In our experiments, these two variables provided enough information to analyze the mechanical response. The heart was stimulated in the right atrium from the spontaneous heart rate [32]. To measure the HR and ICF, a force-displacement transducer (Kent, USA) was attached to the apex of the mouse heart through a 6-0 silk suture (Johnson & Johnson, Brazil) and a low-resistance pulley. The resting tension was adjusted at 0.5 gf. All data were recorded using the Lab-Trax-4 data acquisition interface and subsequently analyzed using Data-Trax2 software with a sample rate of 500 data/s. No significant effect of friction between the silk suture and the pulley was found. The ICF was calculated as the difference between the mean of the maximum and minimum peak values of the recordings. For statistical analysis, data were normalized against the maximum value. Additionally, to calculate the HR value, we obtained the characteristic frequency of power spectra of the recording signals.

To ensure that hearts were in physiological conditions (60–110 mmHg) [33], the coronary perfusion pressure was continuously monitored via a pressure transducer (UFI, Morro Bay, CA) attached to a channel data acquisition recorder (Lab-Trax-4/ 24T, WPI) and a blood pressure display unit (Stoelting, Co., IL, USA).

11.4.2 External electrical stimulation

Hearts were allowed to beat spontaneously (380 beats min^{-1}) (6.33 Hz) for 30 min [32]. Then, a stimulus was applied with two platinum stimulation electrodes (Sigma Aldrich Chemical Co.St., 0.25 mm diameter) connected to a data acquisition card (NI SCB-68). They were superficially attached to the high right atrium, avoiding cardiac tissue damage. The stimulus amplitude was achieved using a function generator developed in LabVIEW; a square wave of 10.0 ms duration with a voltage amplitude of 2 V was applied [33].

11.4.3 Statistical methods

All data are reported as mean ± standard deviation. Spectral analysis was performed using the fast Fourier transform (FFT) [34]. It was implemented in MATLAB® (The MathWorks, Natick, MA, USA) [35].

11.4.4 Mathematical model

The nonautonomous Van der Pol oscillator takes the original function and adds a driving function to give a nonlinear ordinary differential equation of second order of the form

$$\ddot{x} - \alpha(1 - x^2)\dot{x} + \omega^2 x = f(t), \tag{11.1}$$

where α and ω are system parameters and $f(t)$ is an external forcing.

Their time period is determined by a time constant or relaxation time (ϵ) given by $\epsilon = \alpha/\omega$. To write a state model, let us take $z_1 = x$ and $z_2 = \dot{x}$ to obtain

$$\begin{aligned} \dot{z_1} &= z_2, \\ \dot{z_2} &= \alpha(1 - z_1^2)z_2 - \omega^2 z_1 + f(t). \end{aligned} \tag{11.2}$$

This equation corresponds to the situation in which a real heart is driven by a pacemaker and can be solved numerically using the Euler, Taylor series, or Runge–Kutta integration algorithm [35].

11.5 Results

It is known that an isolated heart preserves the ability to perform a rhythmic contraction. If the denervated isolated heart is continuously perfused, the activity of the pacemaker continues to trigger the beats [33]. The Langendorff perfused heart model has proved to be an excellent tool to study, among other properties, the contractile response of the whole heart. In this model, the heart is removed from the body. In the absence of preload and afterload of blood, the heart does not lose its contractile property.

To test the real-time signal acquisition and the accuracy of the experimental setup, a set of measurements of the response to simulated heart contraction was recorded. First, a sinusoidal signal with a frequency similar to that of an isolated mouse heart was generated in LabVIEW (6.33 Hz) [32]. The maximum peak value represents the systole. Second, the experimental setup detects this value, and from there, it generates the first pulse 15 ms later. Then it generates a pulse every 10 ms. T_0 denotes the period of the limit cycle. A stimulus is delivered at a time T_c after the peak value is detected. Thus, a change in the cycle length from T_0 to T_1 is recorded (see Fig. 11.2A.)

Fig. 11.2B shows the real-time signal acquisition results of the response detected to every single pulse applied. Furthermore, to quantitatively characterize the signal acquisition due to stimulus applied, the determination coefficient (R^2) was calculated ($R^2 = 0.9979$).

We next investigated the feasibility of the experimental setup with an isolated mouse heart. Measurements of the contractile response were recorded. A typical mechanical response of a whole heart is illustrated in Fig. 11.3. The mechanical response has a spontaneous rate, 6.33 Hz, approximately [32]. As can be seen, the amplitude of the contractile force was maintained approximately constant (0.2 ± 0.01 gf).

Fig. 11.4 shows our experimental results of the mouse heart response to three single pulses applied following the protocol as mentioned earlier. When the stimulus is applied at the beginning cycle, the length of the next cycle is decreased and the duration of the perturbed cycle is minor to T_0, leading to delay of the phase (Fig. 11.4A&B). Interestingly, an irregular contraction in the second cycle was observed (Fig. 11.4C). Experiments reported in cardiac cells report that the effect is observed in the same cycle when a stimulus is delivered [27]. The durations of the cycles following the perturbed cycle are all very close to T_0.

To evaluate whether the contractile response of the isolated heart corresponds to the dynamics of a relaxation-oscillator, we developed a comparative study between experimental and theoretical PRCs.

Recall that PRC tells us how a stimulus at a phase in the cycle alters the time of the next cycle. We can define the phase of a rhythm by assigning

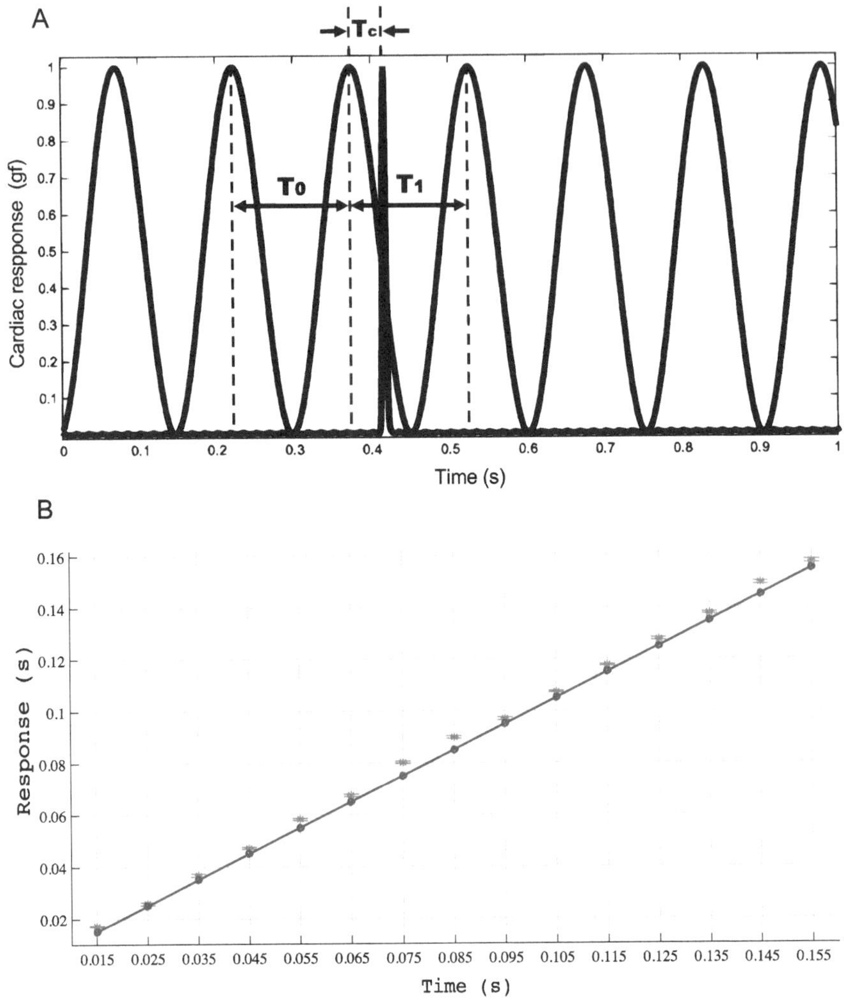

FIGURE 11.2 **Single-pulse protocol.** (A) The protocol for determining the phase response curves. (B) Calibration curve of the experimental setup. The red (gray in print version) line corresponds to the ideal response.

the peak value of the first mechanical response to zero and the phase of the second peak value to be 157 ms. The interspike interval, T_0, is measured between two consecutive peaks ($T_0 = 157$ ms). The stimulus is applied at T_c after a spike is initiated in the mechanical heart response. The next experiment was realized. The first stimulus is a pulse of 10 ms duration and 2 V amplitude delivered at $T_c = 15$ ms. This stimulus induces an immediate perturbed cycle of length T_1. The successive stimuli were applied with increases of 10 ms until a maximum of 155 ms. They are also measured.

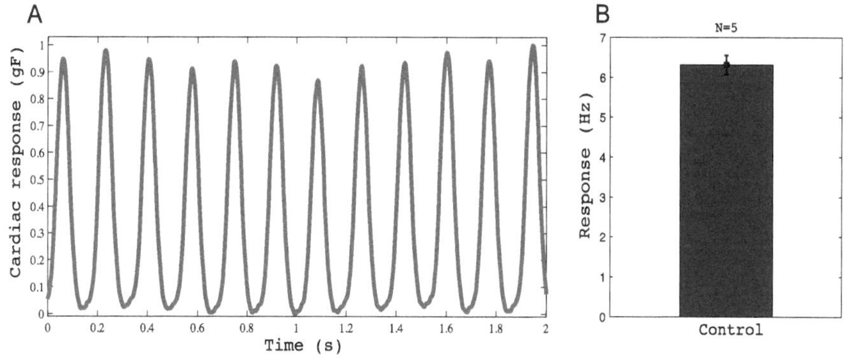

FIGURE 11.3 **Isolated heart response.** (A) Illustrative example of heart contractile response. Normalized force is shown. (B) Natural frequency of oscillation of the isolated heart (6.33 Hz) according to [32].

The PRC was experimentally obtained by means of the phase. The phase, ϕ, is defined as $\phi = T_c/T_0$. As a reference, a phase of zero was assigned to spike initiation, defined as the maximum value of the contractile response. The phase resetting is defined as $\Delta\phi = (T_0 - T_c)/T_0$. Fig. 11.5 illustrates the experimental PRC. Interestingly, multiple transitions between the advance phase and phase delay were observed.

To compare the dynamic response, we solved the Van der Pol oscillator defined by Eq. (11.2) to numerically illustrate the type of PRC for a relaxation-oscillator. The corresponding parameter values are $\alpha = 55$, $\omega = 43.5$ rad/s, $\epsilon = 1.264$. The driving function $f(t)$ corresponds to a square wave signal of 10.0 ms duration with frequency $w_1 = 10$ w and amplitude values in the range [5000, 11000, 20000]. We observe that all PRCs are of Type 1 (see Fig. 11.6). Furthermore, perturbations at phases slightly below $\phi = 0.4$ induce a phase delay, while perturbations applied at phases above cause phase advance. Finally, applying the disturbance to the last 20 ms of the cycle causes a phase delay (Fig. 11.6A&B). When the highest amplitude value in the disturbance was applied, it induced a phase delay for almost the entire cycle (Fig. 11.6C). Phase planes and temporal solutions of the Van der Pol oscillator are not shown.

11.6 Conclusion

The rhythmic activity of cardiac pacemaker cells, pancreatic cells, and neural circuits is a well-known biological example of self-sustaining oscillatory patterns. Phase-resetting theory was developed in limit-cycle oscillators under the assumption that external perturbations affect only the phase, not the oscillation amplitude.

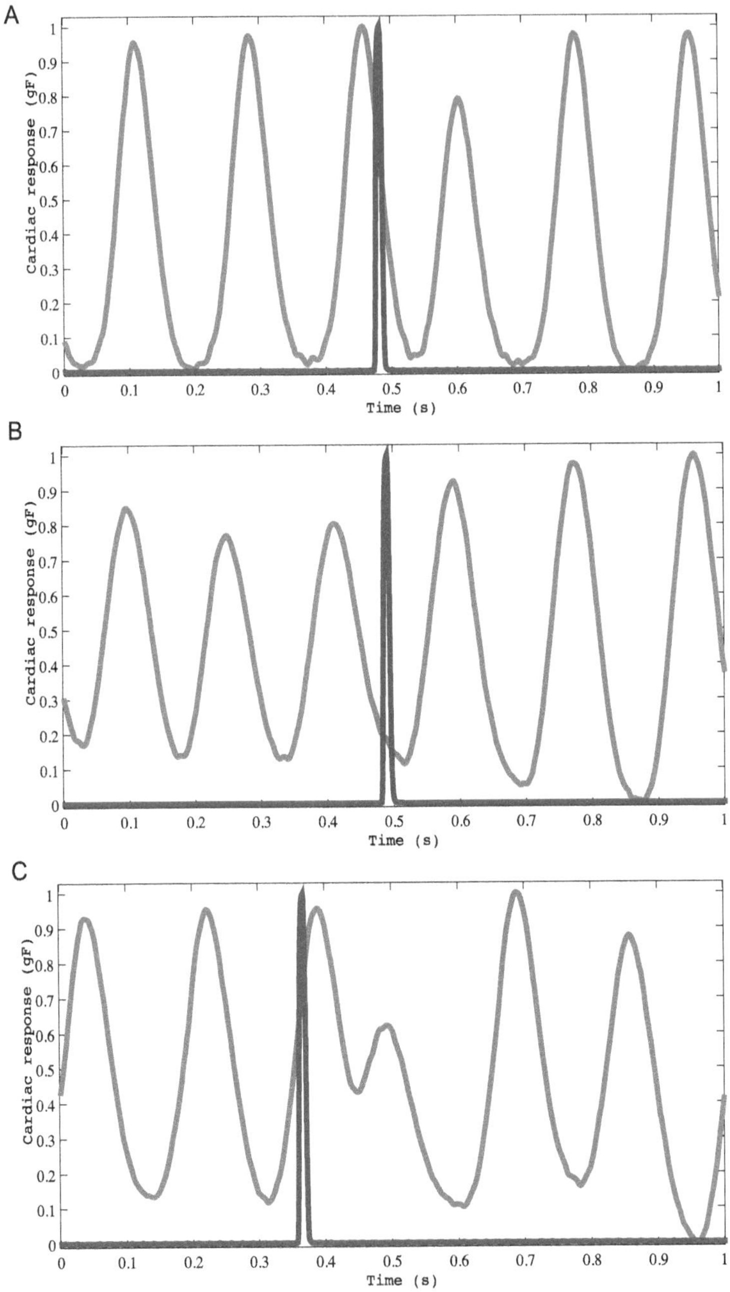

FIGURE 11.4 Experimental response to the single-pulse protocol. The blue (dark gray in print version) line is the contractile heart response. The red (gray in print version) line corresponds to an external stimulation of 2 V amplitude and 10 ms duration applied at (A) 25 ms, (B) 75 ms, and (C) 145 ms.

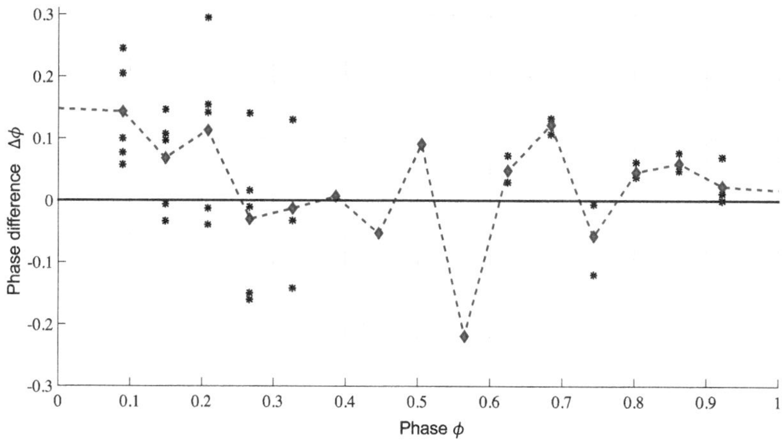

FIGURE 11.5 Experimental phase response curve. The black asterisk represents a stimulus applied in each phase, while the blue/red (dark gray/gray in print version) diamond corresponds to the average of experimental results.

This study measured the phase shifts in response to stimuli delivered at different times in the contractile heart cycle. According to our experimental protocol, the starting point to apply the stimulus corresponds to maximum contraction. Thus, the heart begins relaxation in the initial half cycle. As a result, the applied stimulus affects the contractile response and the dynamics of Ca^{2+}. This could be the cause of amplitude contraction variations. On the other hand, the immediate and subsequent cycles were affected when the stimulus was applied in the final half cycle (Fig. 11.4). Mechanical restitution is the time-dependent recovery of the ability of the heart to contract following the previous beat. A premature stimulus or extra systole introduced soon after the relaxation of a steady-state beat produces an attenuated contraction [36]. In addition, the intracellular Ca^{2+} dynamics could be affected. This is consistent with results reported by Fozzard and Gibbons [37].

To elucidate whether the heart behaves like a relaxation-oscillator when disturbed by an electrical stimulus, we investigated the phase-resetting properties of the mechanical response of the isolated heart. The cardiac conducting system has been studied as a nonlinear oscillator. Earlier studies have documented systematic phase-resetting experiments on biological heart cells. For example, Guevara and colleagues studied phase resetting of spontaneously beating embryonic ventricular heart cell aggregates [38]. This study reports Type 1 PRCs where perturbations begin with a phase advance and finish with a phase delay. Similar PRCs are reported on the phase resetting response of the sinoatrial node [30,39]. Our experimental results reveal that the stimulus applied resets the heart contractile activity phase by delaying or advancing the time of occurrence of the sponta-

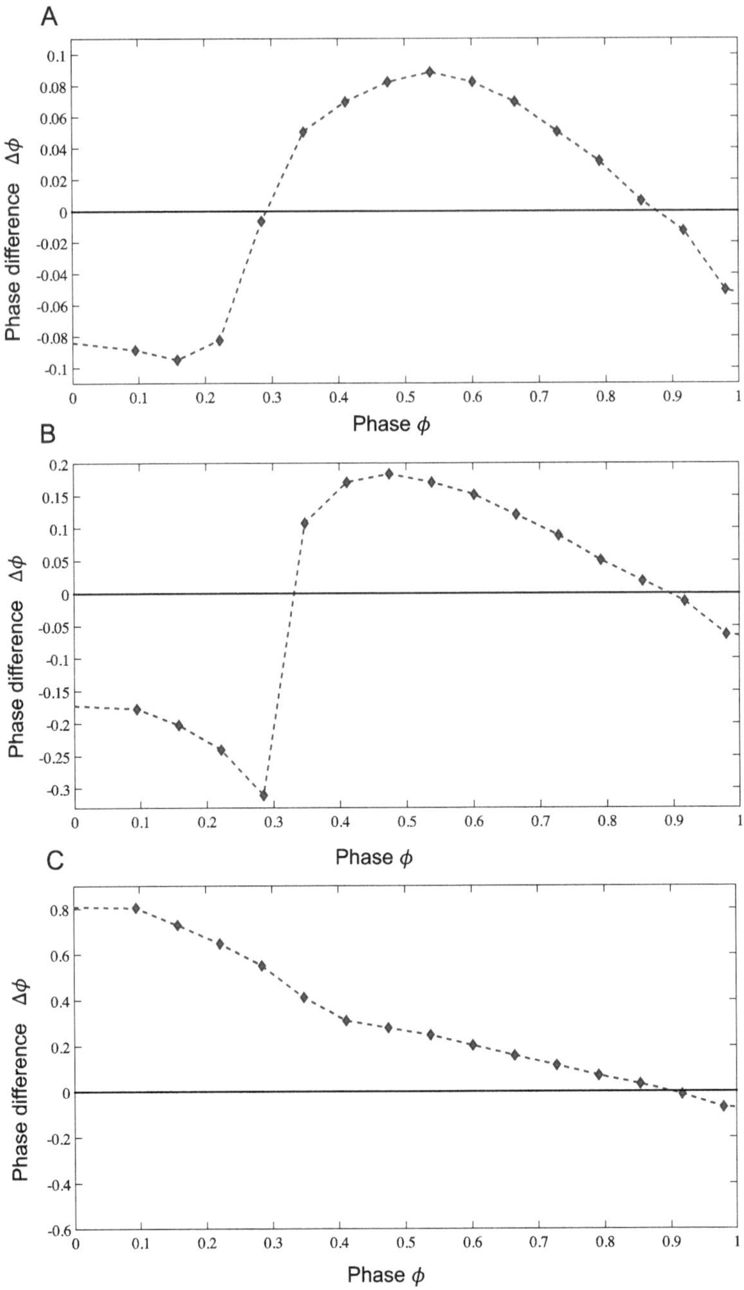

FIGURE 11.6 Theoretical phase response curve. PRCs are obtained solving Eq. (11.2) with parameter values $\alpha = 55$, $\omega = 43.5$ rad/s, $\epsilon = 1.264$. The driving function $f(t)$ corresponds to a square wave signal of 10.0 ms duration and an amplitude value of (A) 5000 ua, (B) 11,000 ua, and (C) 20,000 ua.

neous heartbeat several times. This PRC does not correspond to that of a relaxation-oscillator (Fig. 11.5). We suppose that some stimuli were applied in the refractory period. However, we drew this conclusion with reservations because our experimental setup only allows the contraction force to be measured, and it is known that in this state, the heart does not contract properly, decreasing the afterload volume and, therefore, causing a deficit in the supply of blood at high frequencies, thus reducing the efficiency [40].

Fig. 11.6 shows a set of numerically obtained PRCs. If a PRC is negative (positive), this means that the stimulus pushed the phase back (ahead) and thus delayed (advanced) the time of the next cycle. Continuous dependence of the phase shift $\Delta\phi$ on phase ϕ is observed. It corresponds to PRC Type 1 [27]. By comparing Figs. 11.5 and 11.6, we observed that the experimental PRC could be classified as Type 1 [5]. The phase resetting of biological oscillators by an isolated stimulus delivered at various phases of the cycle was demonstrated in spontaneously beating aggregates of cells derived from embryonic chick heart [26]. Our results differ from those reported by other research groups in cardiac cells [16,38,41,42]. They performed phase restoration studies in cell aggregates and considered the electrical response as an output variable. On the other hand, the ionic current mechanisms that underlie phase resetting properties of embryonic cardiac cells are complex [43]. Our experimental setup allows us to consider the isolated heart as a black box. The output variable is a mechanical force; the ionic currents and molecular events associated with ventricular contraction and relaxation are not monitored. However, the interplay of ionic currents and the PRC is subtle and complex to address with experimental methods. The question arises under what ionic conditions cardiac contraction loses the properties of a relaxation-oscillator. This remains an open question.

Experimental studies of phase resetting require experimental systems with low noise, a comparatively short period, and a sufficiently long experimental duration. Performing slight variations of parameters becomes a burden because of physiological limitations. Any experimental system will be subjected to noise, and in some cases, there can be significant fluctuations in the period of biological oscillators. If the period is not constant, there is necessarily uncertainty in measuring both the phase of the stimulus and the resetting induced by the stimulus. To carry out resetting experiments using a variety of stimulus amplitudes requires very stable oscillations and sufficiently fast oscillations so that repeated resetting experiments can be carried out. We strongly suggest further experiments applying panoramic optical mapping from the entire outer surface of Langendorff-perfused hearts to simultaneously measure the transmembrane voltage and intracellular free calcium levels [44].

Conflict of interest statement

The authors declare that the research was conducted in the absence of any commercial or financial relationships that could be construed as a potential conflict of interest.

Author contributions

BAE and JRG conceived and designed the research. CEGL, AR, and KGC performed the experiments. CEGL and JRG performed the formal analysis. CEGL and JRG wrote the manuscript. JRG supervised the research and revised the manuscript critically. All authors read and approved the final manuscript.

Funding

This research was partially supported by Consejo Nacional de Ciencia y Tecnología (CONACYT, Mexico) under grant CF-MI-2019-116335 and by fellowships to CEGL.

References

[1] L. Glass, Synchronization and rhythmic processes in biology, Nature 410 (2001) 277, https://doi.org/10.1038/35065745.
[2] D.M. Bers, Cardiac excitation-contraction coupling, Nature 415 (2002) 198–205, https://doi.org/10.1038/415198a.
[3] S. Rzeczinski, N.B. Janson, A.G. Balanov, P.V.E. McClintock, Regions of cardiorespiratory synchronization in humans under paced respiration, Physical Review E 66 (2002) 051909, https://doi.org/10.1103/PhysRevE.66.051909.
[4] C.C. Canavier, F.G. Kazanci, A.A. Prinz, Phase resetting curves allow for simple and accurate prediction of robust n:1 phase locking for strongly coupled neural oscillators, Biophysical Journal 97 (1) (2009) 59–73, https://doi.org/10.1016/j.bpj.2009.04.016.
[5] A.T. Winfree, The Geometry of Biological Time, Springer, New York, NY, 1980.
[6] E.N. Best, Null space in the Hodgkin-Huxley equations: a critical test, Biophysical Journal 27 (1) (1979) 87–104, https://doi.org/10.1016/S0006-3495(79)85204-2.
[7] R. Guttman, S. Lewis, J. Rinzel, Control of repetitive firing in squid axon membrane as a model for a neuron oscillator, The Journal of Physiology 305 (1) (1980) 377–395, https://doi.org/10.1113/jphysiol.1980.sp013370.
[8] D. Somers, N. Kopell, Rapid synchronization through fast threshold modulation, Biological Cybernetics 68 (1) (1993) 393–407, https://doi.org/10.1007/BF00198772.
[9] N. Kopell, D. Somers, Anti-phase solutions in relaxation oscillators coupled through excitatory interactions, Journal of Mathematical Biology 33 (1) (1995) 261–280, https://doi.org/10.1007/BF00169564.
[10] D. Somers, N. Kopell, Waves and synchrony in networks of oscillators of relaxation and non-relaxation type, Physica D. Nonlinear Phenomena 89 (1) (1995) 169–183, https://doi.org/10.1016/0167-2789(95)00198-0.

[11] D.A. Eisner, J.L. Caldwell, K. Kistamás, A.W. Trafford, Calcium and excitation-contraction coupling in the heart, Circulation Research 121 (2) (2017) 181–195, https://doi.org/10.1161/CIRCRESAHA.117.310230.

[12] K.E. Barrett, W.F. Ganong, Ganong's Review of Medical Physiology, 18th edition, McGraw-Hill Medical, New York, 2012.

[13] M.L. Lindsey, Z. Kassiri, J.A.I. Virag, L.E. de Castro Brás, M. Scherrer-Crosbie, Guidelines for measuring cardiac physiology in mice, American Journal of Physiology. Heart and Circulatory Physiology 314 (4) (2018) H733–H752, https://doi.org/10.1152/ajpheart.00339.2017.

[14] A.V. Hill, Wave transmission as the basis of nerve activity, The Scientific Monthly 37 (4) (1933) 316–324.

[15] B.V. der Pol Jun D.Sc., Lxxxviii. On "relaxation-oscillations", The London, Edinburgh, and Dublin Philosophical Magazine and Journal of Science 2 (11) (1926) 978–992, https://doi.org/10.1080/14786442608564127.

[16] B. van der Pol D.Sc., J. van der Mark, Frequency Demultiplication, Nature 120 (1) (1927) 363–364, https://doi.org/10.1038/120363a0.

[17] A. Pikovsky, M.G. Rosenblum, J. Kurths, Synchronization, a Universal Concept in Nonlinear Sciences, Cambridge University Press, 2010.

[18] S.H. Strogatz, Nonlinear Dynamics and Chaos: With Applications to Physics, Biology, Chemistry, and Engineering, CRC Press, 2018.

[19] B. Van der Pol, Biological rhythms considered as relaxation oscillations, Acta Medica Scandinavica 103 (S108) (1940) 76–88, https://doi.org/10.1111/j.0954-6820.1940.tb11082.x.

[20] B. van der Pol D.Sc., J. van der Mark, Lxxii. The heartbeat considered as a relaxation oscillation, and an electrical model of the heart, The London, Edinburgh, and Dublin Philosophical Magazine and Journal of Science 6 (38) (1928) 763–775, https://doi.org/10.1080/14786441108564652.

[21] P. Rowat, A. Selverston, Modeling the gastric mill central pattern generator of the lobster with a relaxation-oscillator network, Journal of Neurophysiology 70 (3) (1993) 1030–1053, https://doi.org/10.1152/jn.1993.70.3.1030.

[22] C.B. Tabi, Dynamical analysis of the FitzHugh–Nagumo oscillations through a modified Van der Pol equation with fractional-order derivative term, International Journal of Non-Linear Mechanics 105 (2018) 173–178, https://doi.org/10.1016/j.ijnonlinmec.2018.05.026.

[23] A. Beuter, L. Glass, M. Mackey, M. Titcombe, Nonlinear Dynamics in Physiology and Medicine, Springer, 2010.

[24] R. Isermann, M. Münchhof, Identification of Dynamic Systems: An Introduction with Applications, Springer, 2011.

[25] J. Ang, B. Ingalls, D. McMillen, Chapter ten - probing the input–output behavior of biochemical and genetic systems: system identification methods from control theory, in: M.L. Johnson, L. Brand (Eds.), Computer Methods, Part C, in: Methods in Enzymology, vol. 487, Academic Press, 2011, pp. 279–317.

[26] P.A. Tass, Phase Resetting in Medicine and Biology, 12th edition, Springer-Verlag Berlin Heidelberg, 1999.

[27] A. Granada, R. Hennig, B. Ronacher, A. Kramer, H. Herzel, Chapter 1 phase response curves: elucidating the dynamics of coupled oscillators, in: Computer Methods, Part A, in: Methods in Enzymology, vol. 454, Academic Press, 2009, pp. 1–27.

[28] J. Clay, M. Guevara, A. Shrier, Phase resetting of the rhythmic activity of embryonic heart cell aggregates. experiment and theory, Biophysical Journal 45 (1984) 699–714, https://doi.org/10.1016/S0006-3495(84)84212-5.

[29] K. Grudziński, J.J. Zebrowski, Modeling cardiac pacemakers with relaxation oscillators, Physica A. Statistical Mechanics and its Applications 336 (1) (2004) 153–162, https://doi.org/10.1016/j.physa.2004.01.020.

[30] M.R. Guevara, H.J. Jongsma, Phase resetting in a model of sinoatrial nodal membrane: ionic and topological aspects, American Journal of Physiology. Heart and Circulatory Physiology 258 (3) (1990) H734–H747, https://doi.org/10.1152/ajpheart.1990.258.3.H734.
[31] K. Broadley, The Langendorff heart preparation reappraisal of its role as a research and teaching model for coronary vasoactive drugs, Journal of Pharmacological Methods 2 (2) (1979) 143–156, https://doi.org/10.1016/0160-5402(79)90038-X.
[32] R.M. Bell, M.M. Mocanu, D.M. Yellon, Retrograde heart perfusion: the Langendorff technique of isolated heart perfusion, Journal of Molecular and Cellular Cardiology 50 (6) (2011) 940–950, https://doi.org/10.1016/j.yjmcc.2011.02.018.
[33] F.J. Sutherland, D.J. Hearse, The isolated blood and perfusion fluid perfused heart, Pharmacological Research 41 (6) (2000) 613–627, https://doi.org/10.1006/phrs.1999.0653.
[34] R. Dzwoncyk, R. Neuman, C. Brown, Frequency analysis of the human ecg during ventricular fibrillation, in: Proceedings of the Fifteenth Annual Northeast Bioengineering Conference, 1989, pp. 3–4.
[35] MATLAB, Version 7.10.0 (R2010a), The MathWorks Inc., Natick, Massachusetts, 2010.
[36] C. Dumitrescu, P. Narayan, Y. Cheng, I.R. Efimov, R.A. Altschuld, Phase i and phase ii of short-term mechanical restitution in perfused rat left ventricles, American Journal of Physiology. Heart and Circulatory Physiology 282 (4) (2002) H1311–H1319, https://doi.org/10.1152/ajpheart.00464.2001.
[37] H.A. Fozzard, W.R. Gibbons, Action potential and contraction of heart muscle, American Journal of Cardiology 31 (2) (1973) 182–192, https://doi.org/10.1016/0002-9149(73)91031-X.
[38] M.R. Guevara, A. Shrier, L. Glass, Phase resetting of spontaneously beating embryonic ventricular heart cell aggregates, American Journal of Physiology. Heart and Circulatory Physiology 251 (6) (1986) H1298–H1305, https://doi.org/10.1152/ajpheart.1986.251.6.H1298.
[39] F. Cantini, M. Varanini, A. Macerata, M. Piacenti, M.-A. Morales, R. Balocchi, Cardiac resynchronization therapy and phase resetting of the sinoatrial node: a conjecture, Chaos. An Interdisciplinary Journal of Nonlinear Science 17 (1) (2007) 015106, https://doi.org/10.1063/1.2431079.
[40] J.E. Hall, Guyton and Hall Textbook of Medical Physiology, 12th edition, Saunders/Elsevier, 2011.
[41] L. Glass, Y. Nagai, K. Hall, M. Talajic, S. Nattel, Predicting the entrainment of reentrant cardiac waves using phase resetting curves, Physical Review E 65 (2002) 021908, https://doi.org/10.1103/PhysRevE.65.021908.
[42] M. Guevara, L. Glass, Phase locking, period doubling bifurcations and chaos in a mathematical model of a periodically driven oscillator: a theory for the entrainment of biological oscillators and the generation of cardiac dysrhythmias, Journal of Mathematical Biology 14 (1) (1982) 1–23, https://doi.org/10.1007/BF02154750.
[43] G.B. Ermentrout, R.F. Galan, N.N. Urban, Reliability, synchrony and noise, Trends in Neurosciences 31 (8) (2008) 428–434, https://doi.org/10.1016/j.tins.2008.06.002.
[44] P. Lee, C. Calvo, J. Alfonso-Almazán, J. Quintanilla, F. Chorro, P. Yan, L. Loew, D. Filgueiras-Rama, J. Millet, Low-cost optical mapping systems for panoramic imaging of complex arrhythmias and drug-action in translational heart models, Scientific Reports 7 (1) (2017) 43217, https://doi.org/10.1038/srep43217.

Index

A

Ab cross-reaction, 114
Ab dosing, 47, 63
Absolute convergence abscissa, 49
Acid
  deoxyribonucleic, 6, 175
  nucleic, 6, 175
  ribonucleic, 6, 175
Acute, 4, 11, 12
  infections, 3, 12, 28
    HIV, 152, 165
    viral, 4
Adaptive immune response, 113
Adenovirus, 190
Adjuvant, 176
Administration
  drug, 2, 116
  insulin, 89, 96
American Diabetes Association (ADA), 88
Amount
  drug, 24, 42, 44, 122
  insulin, 79, 89, 96
  virus, 189
Analysis
  input-output, 47
  PTA, 57
  stability, 12, 13
  system, 44
  worst-case, 42, 55
Antigenic shift, 111, 112
Antimicrobials, 3
Antiviral, 13, 23–26, 114
  dose, 24, 32
  drugs, 114
  therapy, 23
  treatment, 22, 23, 25, 27, 28, 36, 37, 124
Approximation dynamic programming (ADP), 117
Artificial equilibrium, 68, 73, 74
Artificial pancreas (AP), 5, 41, 67, 74, 84, 89
Asymptotic stability (AS), 15, 21
  of the equilibrium, 15
Attractive set, 16, 19
Attractivity, 16, 17, 19, 21

B

Basic reproduction number, 14
Bidirectional LSTM, 92, 93
Bidirectional recurrent NNs (BRNNs), 92
Bidirectional structure, 92, 93
BiLSTM, 98, 99, 102, 104
  NN, 98, 102
Biological models, 139
Blood glucose (BG), 68
Blood glucose level (BGL), 96
Boundary
  insulin-on-board, 79
  IOB, 69, 79, 80, 82

C

Cancer cells, 190
Cardiac cells, 209, 215
Cell
  cancer, 190
  cardiac, 209, 215
  effector, 154–156, 159, 161, 165
  human, 115
  immune, 155
  infected, 4, 11, 13, 26, 28, 111, 113, 114, 122, 140, 156, 162, 174, 184, 197
  infection, 112
  latently infected, 153–156, 161, 162, 165, 167, 169
  life, 175
  LSTM, 91–93
  susceptible, 12, 13, 22, 26, 116, 140
  tumor, 6, 7, 189–193, 197, 200
  uninfected, 4, 26, 27, 37, 122

Chemical master equation (CME), 6, 178
Chronic
  disease, 6
  infections, 3
    viral, 6
Clearance, 14, 45–47, 55, 56, 179
  intercompartmental, 45
  parameter, 55, 56
  state, 182, 183
  systemic, 45
  value, 46, 55
  viral, 159, 160, 162, 165
  virus, 184
Coefficients of the polynomial, 132
Continuous glucose monitor (CGM), 5, 67, 89
Control
  algorithm, 89, 90, 99
  elite, 6, 151, 162
  gain, 153, 161
  impulse, 152
  impulsive, 5, 25, 117, 118, 147
    inverse optimal, 118
  infection, 12
  input, 69, 79, 118, 152, 153, 162, 190, 196
  instants, 118
  law, 42, 63, 117–120, 122, 154
  objective, 4, 24, 26–28, 32, 159, 193, 197
  offset-free, 80
  optimal, 12, 36, 117, 118, 153
  problem, 27, 169
  scheme, 84, 116
  sequence, 35
  signal, 96, 153, 161
  single interval, 36
    quasioptimal, 32, 36
  sliding mode, 153, 154, 159, 161, 162, 164, 165, 169
  systems, 67, 117
  theory, 2, 4, 5, 41, 63, 115, 129, 152, 154, 169, 190, 193, 206
  transcriptional, 176
  viral, 152, 154, 165
  virus, 151
Controllability, 152
Cost function, 194
Covariates, 46
COVID-19, 4, 12, 22, 25, 37, 124
Critical value, 14
Cytolytic killing, 154, 161, 162

D

Dataset description, 96
Deep learning, 5
Deep neural networks (DNN), 5, 90
Deoxyribonucleic acid (DNA), 6, 175
Depot compartment, 42, 122
Description
  dataset, 96
  system, 95
Diabetes mellitus (DM), 87
Digital twins, 1
Direct multistep ahead prediction strategy, 94
Discontinuous control
  action, 67
  signal, 6
Disturbance
  estimator, 82
  model, 68–75, 84, 96, 129, 153, 211
DNA replication, 175
DNNs, 5, 90, 104
Domain of attraction (DOA), 15, 16
Dose
  adjustment, 60, 61
  antiviral, 24, 32
  drug, 4, 116
  Goldilocks, 29, 34
    antiviral, 28
  individualized, 57
  insulin, 76–78, 82, 99
  levels, 63
  loading, 54
  maintenance, 44, 54, 56
  tailoring, 5, 116
  viral, 6, 190, 194, 197
Dosing interval, 41, 44, 51–53, 56, 59–63
Dosing regimens, 47, 57
Drug
  administration, 2, 116
  amount, 24, 42, 44, 122
  antiviral, 114

concentrations, 12, 25, 41–44, 63, 115, 121
delivery, 11, 115
development, 1, 115
distribution, 121
doses, 4, 116
dosing, 54
individualized, 47, 58
strategies, 63
effect, 115
efficacy, 41, 121
elimination rate, 24
exposure, 41
infusion rate, 43
intake, 5, 117, 121, 122
therapies, 12
Dynamics
elite control, 158–161
growth, 154
heart, 206
HIV, 152
nonlinear, 205, 206
PK/PD, 3, 121
population, 203
sliding mode, 153, 160
state, 59
viral, 118, 120, 121, 183
virus, 154

E

EBEs, 60, 62
Effect
drug, 115
fixed, 46
random, 47
Effector cells, 154–156, 159, 161, 165
EKF equations, 141
Elite control, 6, 151, 162
dynamics, 158–161
HIV, 151, 152, 154, 156, 158–162, 165, 169
Empirical Bayes estimations (EBE), 60
Epidemic, 110
Equation
chemical master, 6, 178
EKF, 141
filtering, 135, 136, 139, 141, 144
KF, 130, 141
Van der Pol, 133
Equilibrium
artificial, 68, 73, 74
characterization, 15, 68
point, 12, 15–17, 20, 21, 28, 30, 71, 73, 117, 119
set, 12, 15–17
stability, 15
asymptotic, 15
states, 12, 17, 27, 30
trajectory, 56
unstable, 30, 33
Equivalent control, 153, 161, 162, 164
Estimation
insulin-on-board, 79
state, 59, 129–131, 139, 140, 146
polynomial, 140, 144
Extended Kalman filter (EKF), 144
equations, 141
External electrical stimulation, 208

F

Fast Fourier transform (FFT), 208
Feedback
control law, 59
state, 42, 61, 63
Filter, 5
Kalman, 6, 129
extended, 141, 144
Filtering equations, 135, 136, 139, 141, 144
Fixed effect, 46
Food and Drug Administration (FDA), 114
Free virus, 14, 140
Function
cost, 194
polynomial, 132
second-order, 138
state, 79

G

Gene regulation models, 175
General infectious disease system, 139
Glomerular filtration rate (GFR), 46
Glucose
controller, 96
insulin, 89

levels, 87, 88, 96, 98, 100
control, 87, 89
measurement, 89, 97
sensor, 89, 96
therapy, 90
Glycemia, 69, 75–77
Goldilocks antiviral dose (GAD), 28
Goldilocks antiviral treatment, 28
Goldilocks doses, 29, 34

H

HCV
infection, 6, 173, 179, 182
intracellular, 174
replication, 173, 174, 184
Heart, 203–209, 213, 215
contraction, 209
contracts, 204
dynamics, 206
human, 206
isolated, 7, 204, 207, 209, 213, 215
rate, 204, 207
wall, 204
Heartbeat, 7, 203, 206, 207
Hepatitis C virus (HCV), 6, 173
HIV
dynamics, 152
elite control, 151, 152, 154, 156, 158–162, 165, 169
infection, 131, 151, 152, 154–156, 160, 165, 169
acute phase, 152
dynamics, 152
therapy, 6
virions, 154, 155
Host immune response, 113
Human
cells, 115
heart, 206
influenza, 110
Human immunodeficiency virus (HIV), 67, 151
Human leukocyte antigen (HLA), 151
Hyperglycemia, 69, 75–78, 80–82, 84, 87, 89, 100, 104
Hypoglycemia, 74–78, 80–82, 84, 89, 104
safety condition, 68

I

I/O approach, 58
IAV infection, 112, 113
Immune
cell, 155
infiltration, 112
response, 1, 6, 12, 110, 116, 123, 131, 152–154, 156, 159, 189, 192
system, 6, 14, 151, 153, 165, 176, 190, 192
Immunological control objective, 158
Impulsive
control, 5, 25, 117, 118, 147
instants, 118
schemes, 123
strategies, 190
systems, 12, 152, 190
inputs, 67
scheme, 25
state, 118
systems, 25, 67, 68, 73
Impulsive nonlinear MPC (iNMPC), 194
In-host infection, 13
Individualized dose, 57
Individualized drug dosing, 47, 58
Infected cells, 4, 11, 13, 26, 28, 111, 113, 114, 122, 140, 156, 162, 174, 184, 197
Infection
cells, 112
target, 154
chronic, 3
control, 12
HCV, 6, 173, 179, 182
HIV, 131, 151, 152, 154–156, 160, 165, 169
in-host, 13
influenza, 5, 109, 110, 131
rate, 14, 24–26, 159, 161
site, 43, 63
viral, 4, 24, 115, 130, 139, 143, 147, 169, 173, 174, 189
Infectious diseases, 2, 3, 6, 12, 13, 116, 124, 131
Influenza, 109–111, 113, 118, 123, 130
epidemics, 109, 130
human, 110

infection, 5, 109, 110, 131
pandemics, 109, 110
seasonal, 110, 114
strains, 114
system, 122
treatment, 114–116, 122, 123, 191
tailoring, 120, 121
viral genome, 111
virions, 114
virus, 109, 111, 112, 115, 123
infection, 112, 114
pathogenesis, 112
Infusion, 41, 44, 57, 153
duration, 41, 53, 63
insulin, 68, 89
Inherit topology, 13
iNMPC, 194, 196–198
Innate immune response, 174, 178, 179, 184
Innate immune system, 176
Input-output analysis, 47
Input-output formula, 54, 55
Insulin, 75, 79, 82, 87, 88, 100
action, 69, 79
administration, 89, 96
amount, 79, 89, 96
boluses, 74, 79
doses, 76–78, 82, 99
glucose, 89
infusion, 68, 89
injections, 68, 88
input, 80
pump, 74, 80, 96
resistance, 87
secretion, 88
sensitivity, 75
stacking, 68, 78, 84
Insulin-on-board (IOB), 68
constraint, 78
estimation, 79
Intercompartmental clearance, 45
Inverse optimal control, 5
Inverse optimal impulsive control, 118
law, 119, 120
IOB boundary, 69, 79, 80, 82
ISG
overexpression, 182
protein, 173, 174, 178, 183, 184
Isolated heart, 7, 204, 207, 209, 213, 215
Itô process, 132
stochastic differential, 132

K

Kalman filter (KF), 6, 129
equations, 130, 141
Killing
action, 159, 161, 162, 165
efficacy, 151, 156, 159, 162
process, 152, 154–156, 165
rate, 156

L

Langendorff model, 209
Langendorff system, 207
Laplace transform, 49
Latently infected cells, 153–156, 161, 162, 165, 167, 169
Left ventricle (LV), 204
Loading dose, 54
Long short-term memory (LSTM), 91
recurrent neural network, 91
LSTM, 91, 98, 99, 102, 104
cell, 91–93
memory, 92
network, 91
NN, 98
Lyapunov's stability theorem, 17

M

Maintenance dose, 44, 54, 56
Mathematical modeling, 1, 41, 115, 152, 174, 177, 191
Mathematical models, 1, 4, 5, 11, 13, 116, 117, 131, 139, 152, 174, 178, 184, 190, 206, 208
deterministic, 177
Mechanical response, 204, 207, 209, 210, 213
Meropenem, 42, 44, 45, 47, 54, 57, 63
Minimal inhibitory concentration (MIC), 54
Minimed Real-Time Paradigm, 95
MLP, 93, 98, 102
Model
biological, 139
gene regulation, 175
Langendorff, 209

mathematical, 1, 4, 5, 11, 13, 116, 117, 131, 139, 152, 174, 178, 184, 190, 206, 208
of meropenem, 45
population pharmacokinetic, 42
prediction, 72–74, 77, 80, 84, 191
stochastic, 178
target cell, 139
limited, 4, 12, 13, 152
virus immune response, 139, 143, 147
Model predictive control (MPC), 68, 153, 190
for oncolytic virus therapy, 193
Monte Carlo simulation, 4
mRNA, 174–177, 182, 184
molecule, 175
overexpression, 184
therapies, 177
transcription, 178
treatments, 177
vaccines, 175–177
for infectious diseases, 176
MSA prediction, 89–91, 95, 102, 104
Multilayer networks, 93
Multilayer perceptron (MLP), 93

N

NA inhibitors (NI), 114
Natural killer (NK), 113
Networks
deep neural, 5, 90
long short-term memory recurrent neural, 91
multilayer, 93
neuronal, 96
recurrent, 91
Neural network (NN), 89
Neuraminidase inhibitors (NI), 114
Neuronal network, 96
Nonlinear
dynamics, 205, 206
MPC, 191, 197, 200
strategy, 196
oscillator, 207, 213
polynomial systems, 130
Normoglycemia, 74, 78, 82
zone, 74
Nucleic acids, 6, 175

O

Offset-free control, 80
strategy, 73
Offset-free model predictive control (OF-MPC), 5
Offset-free strategy, 80
MPC, 72
Oncolytic
adenovirus therapy, 6
virus, 6, 189–191
therapy, 6, 67, 189, 191, 193
Optimal control, 12, 36, 117, 118, 153
inverse, 5
laws, 117, 119
strategy, 153
Optimizing influenza therapies, 115
Ordinary differential equations (ODE), 11, 152
Overexpression, 182, 183
factor, 182
ISG, 182
mRNA, 184

P

Pandemic, 110
Paradigm
reachability, 6, 153, 154, 158, 169
robust control, 155
sliding mode control, 169
Pathogens, 3, 131
dynamics, 11
Persistence state, 182
Personalized medicine, 1, 2, 4, 5, 7
Personalized treatment, 118, 123
Pharmacodynamic (PD), 3, 12, 115
Pharmacokinetic (PK), 3, 12, 115
Phase resetting, 204, 211, 213, 215
Phase resetting curve (PRC), 203
Physiological variations, 69, 82
PK parameters, 43, 47, 55, 56
PK/PD dynamics, 3, 121
PK/PD schemes, 115
Plasma concentration, 44, 45, 53, 54, 59, 60, 63
Plateau, 51
Polynomial function, 132
Polynomial state estimation, 144
Population
dynamics, 203

level, 42, 47, 59, 63
median, 76, 82
pharmacokinetic model, 41, 42
PRCs, 206, 207, 213
Prediction
horizon, 90, 94, 95, 98, 102, 194, 196
model, 72–74, 77, 80, 84, 191
MSA, 89–91, 95, 102, 104
Probability of target attainment (PTA), 42
Problem
control, 27, 169
state estimation, 130, 131
Process
Itô, 132
killing, 152, 154–156, 165
stochastic, 131, 177
transcription, 175
translation, 175
Wiener, 131
Productively infected cells, 140, 154, 155, 162
killing process, 155
population, 158, 159
dynamics, 161
Protein overexpression, 183
PTA analysis, 57

Q
Quasioptimal control action, 36
Quasioptimal single interval, 32
control, 32, 36
action, 29
treatment, 29, 31, 32, 34
antiviral, 29, 34, 37
Quasisteady state, 29

R
Random effect, 47
Reachability condition, 153, 159–161, 165
sliding mode, 6
Reachability paradigm, 6, 153, 154, 158, 169
Rebound, 29
virus, 26, 32, 37
Recurrent neuronal networks, 91
Region of attraction, 16
Release factor, 176
Reproduction number, 21, 22, 24, 26, 27
Response
immune, 1, 6, 12, 110, 116, 123, 131, 152–154, 156, 159, 189, 192
adaptive, 113
host, 113
Ribonucleic acid (RNA), 6, 175
Robust control paradigm, 155

S
Safety layer (SL), 68, 78, 80
Scheme
control, 84, 116
impulsive, 123
impulsive, 25
PK/PD, 115
Seasonal influenza, 110, 114
Second-order polynomial function, 138
Sensor-augmented pump (SAP) therapy, 89
Short-term treatment, 34
Signal
control, 96, 153, 161
discontinuous, 6
switching, 80
Simulation
Monte Carlo, 4
scenario, 75, 76, 78, 80, 82, 84
Single interval, 32, 35
control, 36
actions, 32
drug delivery, 12
treatment, 27, 29
antiviral, 27, 29
Sinoatrial node, 7, 203, 207, 213
Sliding manifolds, 158, 160
Sliding mode, 6, 153, 154, 159, 160, 162, 165, 167, 169
control, 153, 154, 159, 161, 162, 164, 165, 169
paradigm, 169
system, 153
theory, 153–155, 160–162, 169
controllers, 153, 154, 160, 164, 169
dynamics, 153, 160
reachability condition, 6
Sliding motion, 153, 160, 161, 164, 166
Soft long-term treatment, 33

Softer antiviral treatment, 30
Stability
analysis, 12, 13
asymptotic, 15, 21
equilibrium, 15
theory, 16
Stabilizing feedback control law, 5, 117
Stable equilibrium point, 34
Stable equilibrium set, 17, 30
Stable steady states, 37
Standard deviation (SD), 77
State
clearance, 182, 183
constraints, 194, 196
dynamics, 59
equilibrium, 12, 17, 27, 30
estimation, 59, 129–131, 139, 140, 146
methods, 129
polynomial, 140, 144
problem, 130, 131
estimator, 59, 60, 141
feedback, 42, 61, 63
function, 79
impulsive, 118
persistence, 182
quasisteady, 29
stable steady, 37
trajectory, 55, 56, 59
variables, 45, 75, 118, 192, 206
vector, 14, 43, 119, 122, 133, 134, 141
State-space, 42
Statistical methods, 208
Steady output, 51
Steady state, 51
Stochastic differential of the Itô process, 132
Stochastic model, 178
Stochastic process, 131, 177
Stochastic simulation algorithm (SSA), 178
Strategy
direct multistep ahead prediction, 94
nonlinear MPC, 196
offset-free, 80
control, 73
MPC, 72
optimal control, 153
Susceptible cells, 12, 13, 22, 26, 116, 140
Switching signal, 80
System
analysis, 44
biology, 115, 178
control, 67, 117
description, 95
general infectious disease, 139
immune, 6, 14, 151, 153, 165, 176, 190, 192
impulsive, 25, 67, 68, 73
Langendorff, 207
target cell, 140
Systemic clearance, 45

T

T1DM treatment, 74
Tailor therapies, 3, 5, 26
Target
cell, 4, 12, 139, 155, 156, 165
growth dynamics, 154
infection, 154
limited model, 4, 12, 13, 152
model, 139, 142, 143
system, 140, 147
concentration, 54, 56
range, 68
set, 73, 75
zone, 76, 77
Theory
control, 2, 4, 5, 41, 63, 115, 129, 152, 154, 169, 190, 193, 206
stability, 16
Therapeutic target, 200
Therapy, 12, 190
antiviral, 23
glucose, 90
HIV, 6
mRNA, 177
oncolytic virus, 6, 67, 189, 191, 193
optimization, 114
optimizing influenza, 115
sensor-augmented pump, 89
tailor, 3, 5, 26
viral, 193
Total amount of virus, 28
Transcription process, 175
Transcriptional control, 176
Translation process, 175

Treatment
 antiviral, 22, 23, 25, 27, 28, 36, 37, 124
  Goldilocks, 28
 influenza, 114–116, 122, 123, 191
 mRNA, 177
 personalized, 118, 123
 quasioptimal single interval, 29, 31, 32, 34
  antiviral, 29, 34, 37
 short-term, 34
 single interval, 27, 29
  antiviral, 27, 29
 soft long-term, 33
 T1DM, 74
 tailoring, 116
  influenza, 120, 121
 two-step, 35
Tumor cells, 6, 7, 189–193, 197, 200
 dynamics, 191
Two-step treatment, 35
Type 1 diabetes mellitus (T1DM), 67

U

Uninfected cells, 4, 26, 27, 37, 122
Unstable equilibrium, 30, 33

V

Vaccines
 for infectious diseases, 175, 177
 mRNA, 175–177
Value
 clearance, 46, 55
 critical, 14
Van der Pol equation, 133
Variability
 interindividual, 41
 unexplained, 46, 47
Viral
 clearance, 159, 160, 162, 165
 control, 152, 154, 165
 doses, 6, 190, 194, 197
 dynamics, 118, 120, 121, 183
 infection, 4, 24, 115, 130, 139, 143, 147, 169, 173, 174, 189
 RNA, 111, 112, 174, 184
 therapy, 193
Virion, 111–114
Virtual patient, 1, 22, 25, 62
Virus
 clearance, 184
 control, 151
 dynamics, 154
 effectiveness, 189
 engineering, 190
 immune response model, 139, 143, 147
 influenza, 109, 111, 112, 115, 123
 injection, 192
 load, 26, 190
 oncolytic, 6, 189–191
 particles, 4, 192, 197
 peak, 4, 26, 31, 36
 rebound, 26, 32, 37
 total amount, 28
Volume of distribution, 44

W

Wiener process, 131
World Health Organization (WHO), 3, 114
Worst-case
 analysis, 42, 55
 realization, 56

Z

Zone MPC (ZMPC), 68, 72, 75–77

FSC
www.fsc.org
MIX
Papier | Fördert gute Waldnutzung
FSC® C083411

Kolibri 360 GmbH
Ferdinand-Jühlke-Straße 7
99095 Erfurt, Deutschland
produktsicherheit@kolibri360.de